> What I hear I forget,
> What I see I remember,
> What I touch I understand.
>
> -Confucius (555 - 479 BC)

Each day is like the stitch in the fabric of our lives. Over the years, days bring different colors and change patterns. In knitting or crocheting, there is always an option to repeat the pattern we enjoy or to unravel the piece we dislike and start it all over again—with different stitches, different colors, choosing a different pattern. For our own story, all we can do is follow the thread to find where it begins.

In early September 1954, trees along the canals of Amsterdam started to get some fall colors in their leaves. The city was still busy healing World War II wounds, but some people were rushing in the streets of Amsterdam with suitcases and serious faces. The International Congress of Mathematicians (ICM) was once again held in Europe.[1]

Several prominent mathematicians were invited to give plenary talks in Amsterdam, but the most interest seemed to be centered on astoundingly creative John von Neumann (1903–1957), who was invited to give the lecture "Unsolved Problems in Mathematics." He was expected to summarize what had been accomplished since 1900, when David Hilbert (the most famous mathematician at that time) had talked about unsolved problems in mathematics, and also what new problems were challenging mathematicians after half a century. Von Neumann mostly talked about problems in which he was interested, because, as he said:

> The total subject of mathematics is clearly too broad for any of us. I do not think that any mathematician since Gauss has covered it uniformly and fully, even Hilbert did not, and all of us are of considerably lesser width (quite apart from the question of depth) than Hilbert.[2]

Geometry was of special interest in this congress because it had been 100 years since Georg Friedrich Bernhard Riemann (1826–1866) gave his famous inaugural lecture at the University of Göttingen, "On the Hypotheses which Lie in the Foundation of Geometry." One of the problems that later grew out of that 1854 talk was: Is there any surface in 3-space that has constant negative curvature? Such a surface is called a *hyperbolic plane*. Hilbert answered this question in 1901, by proving that one cannot have an equation describing such a surface. (More about it will be discussed in Chapter 1.) However, the crocheting construction of such surfaces (hyperbolic planes) is the subject of this book. I will talk about this history in more detail in Chapter 4; however, in 1954, there were several other connections to the story of this book.

For example, at the congress, a young graduate student who later became a well-known mathematician, Robert Osserman, presented a talk related to a problem about embedding a hyperbolic plane in 3-space.[3]

This problem was suggested to Osserman by his thesis advisor, Lars Ahlfors (1907–1996), who was one of the first recipients of the Fields Medal. There is an interesting story about it that Robert Osserman told me.

In order to celebrate the 100th anniversary of the birth of the Riemann surface in Riemann's doctoral dissertation, a conference was held at Princeton in December 1951. Osserman was a graduate student at the time and had recently begun his doctoral studies at Harvard under the supervision of Ahlfors. Ahlfors was to give the opening address at the conference, and he suggested that his graduate student Osserman attend also. Prior to the opening talk, Lipman Bers (1914–1993) asked if he could present the assembled experts with two open problems concerning ordinary surfaces in Euclidean 3-space regarded as Riemannian surfaces. Lipman Bers, who became president of the American Mathematical Society (1975–1976), was born in Riga, Latvia (the same city where I was born in 1954). Osserman tried to crack one of these problems, but he got his breakthrough only in 1954 while in Paris continuing his studies. He wrote to me in 2008:

> Instead of trying to construct a surface in the form of a graph that I could prove had to be hyperbolic, I could try starting with a classical Riemann surface given as a branched covering of the plane that I knew to be hyperbolic, and then see if I could "unravel" it so that it could be embedded as a graph over the whole plane. *I wrote to Ahlfors to explain my "unraveling" process, but he wasn't able to understand my description, so I made a model* using a neighborhood of a simple branch point connecting two sheets; one could cut it along a ray from the branch point, fold it up into a fan shape consisting of 16 wedge-shaped sectors with a 45-degree angle each, and then glue the two sides of the cut back together, obtaining an isometric image of the original surface, with its 720-degree total angle at the branch point. *I mailed the model to Ahlfors, and finally got his affirmation that my (very geometric) argument was valid.* It turned out to be a bit harder than I had anticipated to find a branched surface known to be hyperbolic that could be unraveled in its entirety in this fashion and displayed as a graph over the whole plane, but I was able to do so, and that was the subject of my talk at Amsterdam whose abstract was published in the proceedings of the International Congress. [The emphasis is mine.]

Osserman's result is connected to hyperbolic planes because, in 1871, Henri Poincaré mapped the hyperbolic plane to a disk without changing angles—this mapping is called the *Poincaré model*.

The Poincaré model of the hyperbolic plane was used by some mathematicians, but after this particular ICM it became extremely popular because of a Dutch artist, M. C. Escher (1898–1972). As a part of the 1954 ICM cultural program, there was a special exhibition organized in the Stedelijk Museum, in which the shy Dutch artist showed many fascinating prints and carved wooden balls, as well as some hand-drawn and colored drawings of periodic tessellations of reptiles, birds, and other interlocked creatures.

M. C. Escher, *Reptiles*, Litograph, 1943.

Although Escher did not have any mathematical training, he had an amazing understanding of mathematics, both visually and intuitively. The mathematical influence emerged in his works after 1936, especially after his visit to the Alhambra (discussed in Chapter 3), but Escher was very nervous about having an exhibit at a Congress of Mathematicians. Dutch mathematician N. G. de Bruijn wrote in the catalog:

> In view of the fact that Mr. Escher's work may be said to be a point of contact between art and mathematics, the Organizing Committee ... took the initiative to inaugurating this exhibition. Probably mathematicians will not only be interested in the geometric motifs; the same playfulness which constantly appears in mathematics in general and which, to great many mathematicians is the peculiar charm of their subject, will be a more important element.[4]

Escher's work was extremely well received, and this exhibit was the beginning of a long-term friendship between the famous geometer H. S. M. Coxeter and the artist. Coxeter asked Escher's permission to use some of his work as illustrations in his mathematical papers—a tradition he continued for years, promoting the popularity of Escher's work among the mathematical community. Later Escher in his turn asked Coxeter a question that puzzled him from the artistic point of view: Was there any way he could map the whole plane in a circle? Coxeter knew the mathematical answer to this question and introduced Escher to the Poincaré disk model of the hyperbolic plane, and for many years Escher's famous woodcuts *Circle Limit I, II, III,* and *IV* became iconic representations of the hyperbolic plane.

H. S. M. Coxeter's tiling of Poincaré disk.

M. C. Escher, *Circle Limit I*, 1958, woodcut.

In addition, the young Dutch mathematician Nicolaas Kuiper was apparently thinking in 1954 about the problem of a hyperbolic surface because in mid-1955 he published a proof that there must exist a hyperbolic surface (not defined by equations), but he was not able to produce a concrete construction. (This history will be discussed more in Chapter 1.)

In 1954, Sir Roger Penrose was only a second-year graduate student when he attended the ICM. Somebody suggested that he go and see Escher's exhibit. Escher's work inspired Penrose to try to draw something impossible himself, and he came up with the discovery of an impossible triangle. In 1958, a friend showed the Swedish artist Oscar Reutersvärd the now classical article on impossible figures by Lionel Penrose and Roger Penrose, which featured the Penroses' versions of the impossible triangle and impossible staircase.[5] Reutersvärd then recollected that he already drew a similar object 20 years earlier while doodling in a Latin grammar notebook during class. The Penroses' article prompted Reutersvärd to see the impossible figures as new and worthy as a serious subject to explore, and he became well known as the "father of impossible figures."[6]

Aesthetics have always been important to Penrose. In his famous book, *The Emperor's New Mind*, he wrote:

> My impression is that the strong conviction of the *validity* of a flash of inspiration (not 100 per cent reliable, I should add, but at least far more reliable than just a chance) is very closely bound up with its aesthetic qualities. A beautiful idea has a much greater chance of being a correct idea than an ugly one. At least that has been my own experience.[7]

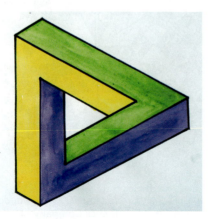

Penrose's impossible triangle.

Reutersvärd's impossible figure.

In September 1954, I was just a few-weeks-old baby in Riga, Latvia, and had no idea about all these connections or the fact that I would become a mathematician, that I would work on the other side of the world, and that the hyperbolic plane would introduce me not only to many extraordinary mathematicians but also to the fiber arts and the many wonderful people I have met because of these crocheted hyperbolic planes.

For me, adventures with the hyperbolic plane were unpredictable. I first learned about a paper model of a hyperbolic plane in June 1997. Professor David Henderson, now my husband, was leading a workshop about teaching geometry to university professors and was showing the model that you can see in the picture below. He said that he learned how to make this model from Bill Thurston in the late 1970s.

That fall, I was scheduled to teach a geometry class at Cornell and therefore was seriously thinking about how to better explain hyperbolic geometry to my students. We were spending summer vacation on a tree farm in Pennsylvania, and while watching my two little girls learning to swim, I was crocheting my classroom set of models of the hyperbolic plane. It turned out to be a success. Students really liked the tactile way of exploring hyperbolic geometry. They told me that after playing with these models, the acquired experiences helped them to move on. Later this set of models was used in several classes and many workshops (and still is used today and is in good shape!).

I learned about the hyperbolic plane from this paper model.

Crocheted hyperbolic plane models are often used in our workshops for school teachers (Washington, DC, summer 2006).

David and I decided to publish a pattern of how to crochet a model of the hyperbolic plane and also to tell about our own discovery of a formula for hyperbolic area—which arose after we played with these models ourselves. We convinced the editor of *The Mathematical Intelligencer* that the idea of using crochet to visualize mathematical concepts was worth publishing and that somebody else might use it too.[8]

Actually, we were right: a few years later there was another crochet model on a cover of *The Mathematical Intelligencer*—this time two mathematicians from Bristol University, UK, Hinke Osinga and Bernd Krauskopf, used crochet technique to show what the Lorenz manifold (from the area of mathematics called dynamical systems) looks like. They wrote that our paper inspired them to use crochet as a medium.[9]

The two "cover girls" finally met in 2008.

Crocheted models featured in *The Mathematical Intelligencer* 2001 (left) and 2004 (right).

The Los Angeles–based Institute For Figuring, which is popularizing the aesthetic qualities of science, invited David and me for our first lecture about the hyperbolic plane to the general public. It generated a lot of interest and was very well received. Our audience was mostly creative and talented people who in their everyday activities were far from mathematics but who had great interest in different geometries and space. This, our initial lecture, grew into a larger interview with Margaret Wertheim.[10] It became my introduction to the art world. I was really surprised when I started to get invitations to participate in fiber art exhibits. The first of them was a group show entitled "Not the Knitting You Know" at Eleven Eleven Gallery, Washington, DC, in spring 2005. That in turn pushed me to look at my crocheted models from a different perspective.

The idea to write this book grew from countless questions I have received after my lectures, during workshops, and over electronic mail. In this book, I am trying to explain to crocheters the mathematics that can be experienced through these models. With weathered mathematicians, I am sharing a tactile way of exploring mathematical ideas. For all others, I hope this book will convey a different way to view mathematics.

Chapter 1

What Is the Hyperbolic Plane? Can We Crochet It?

Positive and Negative Curvature

When we talk about numbers, we can divide them into positive numbers, negative numbers, and zero.

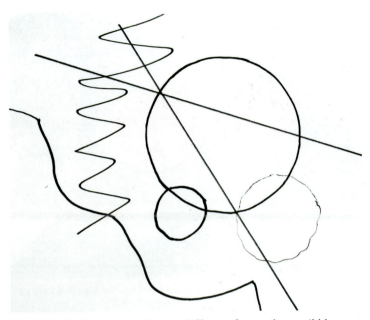

Straight lines and circles are different from other scribbles because they are the same everywhere.

Elevator buttons in France: the ground floor is numbered 0.

When we scribble different lines with a pencil on a piece of paper, we can get different shapes in the plane. The ones that are most useful in geometry are the ones that are "the same everywhere." In the plane such shapes are straight lines and circles.

A straight line is called straight because it is not curved—in other words, we can say it has *zero curvature*. Circles are curved everywhere in the same way—so they have *constant curvature*. The larger the radius of a circle, the less curved it is. For example, if you look at part of a circle with a very large radius, the line will seem straight. Mathematicians express this property by saying that circles with radius R have curvature $1/R$. Imagine that we are sketching a landscape as children first learn to do, just with different curves; there can be hills and there can be valleys in this landscape. In our example we are avoiding cliffs and sharp peaks or valleys. The skyline of the landscape is not uniformly curved everywhere, but at each point we

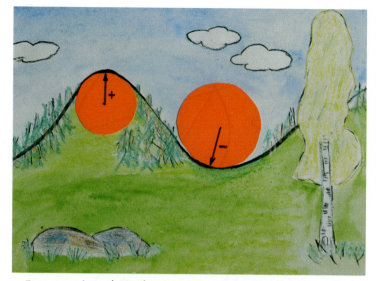

For curves in a plane, the curvature can be negative (in a valley) or positive (on the top of a hill).

Two surfaces with a positive curvature: on the egg it is varying, but on the orange it is (almost) constant.

can measure the curvature by the circle that would "fit" in the curve at that point, as you can see in the picture. To distinguish numerically where the hills are and where the valleys are, we use a notion of one-dimensional positive and negative curvature. We can decide that hills have positive curvature where the circle would be inside the hill, below the ground; for valleys we say that the curvature is negative where the circle is above the ground. At some place positive curvature will turn into negative curvature, and at that place the curvature will be zero. It will also be zero everywhere that the curve is straight.

Can we make a similar distinction with surfaces?

Curvature is a mathematical notion widely used in differential geometry. What does it mean in simple words? If you look, for example, at the surface of your desk or at the floor in your room, you will notice that it is flat—there is no curvature, or we say that the curvature is *zero*. If you look at an orange or an egg, you will see that it is curved "outward"—we say that the curvature is *positive*. The egg is curved more at its tips and curved less in between and thus does not have constant positive curvature, but the orange has almost constant positive curvature.

Now look at the surface of the pear. Most of its surface has positive curvature like the orange or the egg, but there are some points where the curvature is different. How can we describe this difference?

Many leaves have interesting surfaces, like in the picture with kale leaves.

Not all points on the pear have positive curvature.

Kale leaves have negative curvature.

In the early nineteenth century, one of the greatest mathematicians of all times, Carl Friedrich Gauss (1777–1855), explored the idea that surfaces can be distinguished by their *curvature* at different points, which can be positive, negative, or zero. In the 1820s, Gauss was a professional surveyor. This work inspired him to study the intrinsic geometry of surfaces. His concern was how one can determine the curvature of an arbitrary surface without knowing anything about how this surface might be embedded in space—in other words, asking whether and how one could determine the curvature of a surface through measurements made only along this surface, without knowing anything about the shape of this surface. Consider an ant crawling on a big sphere, where the ant cannot see that it is on a sphere; how could this ant distinguish whether it is on a flat surface or on a curved surface? If the ant crawls on a straight line or on any other one-dimensional curve, it can move only in two directions: forward or backward. If the ant is on a two-dimensional surface, then it can choose to go backward or forward but also right or left. Still, the ant cannot see the surface it is on from the outside; it can explore the surface only intrinsically, not leaving this surface.

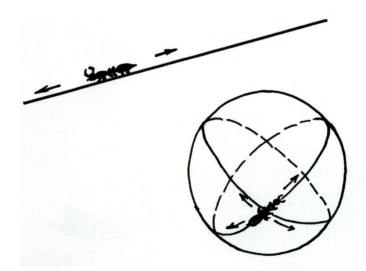

An ant on a curve can go in only two directions, but on a surface there are more choices.

Another great mathematician, Leonard Euler (1707–1783), had already introduced a notion of surface curvature, which was used in eighteenth-century calculus. But to use Euler's method for surfaces, you had to know how the surface is embedded in space. This means that you had to be able to see the surface from another dimension. Gauss was able to prove that it is possible to find a way to determine the curvature of a surface that depends only on intrinsic properties of the surface.[1]

If a surface has constant positive curvature, it will become closed, and it is called a sphere. Gauss used the radius of the sphere to determine the magnitude of the curvature. Every point on a sphere is the intersection of two great circles with the radius R. As defined earlier, the curvature of each of those circles is $1/R$, so Gauss measured the curvature of a sphere as the product of the curvatures of these two great circles. Therefore, he defined the curvature of a sphere with radius R as the quantity $1/R^2$; so, as the radius, R, gets larger, the curvature, $1/R^2$, gets smaller.

Let us now look at a sketch of a three-dimensional landscape. There are hills, passes, and the bowl-shaped bottom of the bay. We can draw two intersecting curves to define a point on the top of a hill and a point in the bay. Then, surface curvature at each of these points will be the product of curvatures of the two intersecting curves. Therefore, in both cases—on the top of a hill and in the bottom of the bay—we have positive curvature, since positive times positive equals positive and negative times negative equals positive. But at the passes, where there is one positive curve and one negative curve, the surface curvature will be negative. In general, we will define the curvature of the surface at a point in terms of the one-dimensional curvatures of two curves on the surface that intersect at that point.

Spheres with different radii: the tennis ball has radius 3.5 cm, and the metal ball has radius 0.5 cm.

Positive curvature is on the top of a hill
(as a product of two positive curvatures)
and in the bay (as a product of two negative curvatures),
but on a pass the curvature is negative
(as a product of positive and negative curvatures).

What Surface Will Have Constant Negative Curvature?

Cut from paper some regular hexagons. Try to put them next to each other. You can see that around each hexagon you can place six other hexagons, and then at each vertex there will be three hexagons and they will lay flat on the table. If you continue adding more hexagons, it looks like the surface of a honeycomb.

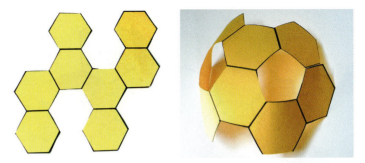

A. Some of the hexagons are removed, so that at each vertex there are only two hexagons.
B. Gluing the hexagons so that there is a pentagonal hole surrounded by five hexagons causes the surface to bend.

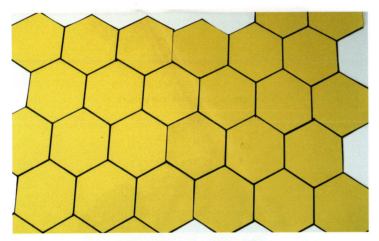

Regular hexagons tile the plane without gaps or overlaps.

Now take a pentagon with the same side lengths as the hexagons you had before and surround that with five hexagons. Notice that this surface is no longer flat. If you continue doing this (see the figures labeled A, B, and C) then you will get a polyhedron that is sometimes called a truncated icosahedron or an approximation of a soccer ball (in most of the world this is called a football).

C. Continue gluing so that hexagons surround a pentagonal hole, and eventually the surface will close.
Five hexagons around a pentagon make a soccer ball (football) with approximately constant positive curvature.

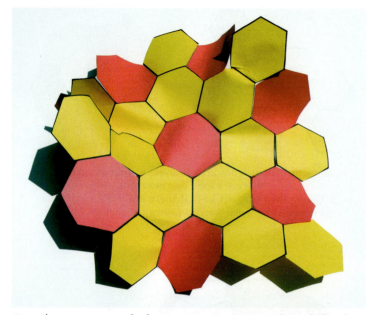

Seven hexagons around a heptagon approximates a hyperbolic plane (constant negative curvature).

Flattening positive curvature.

Finally, see what happens if instead you surround with hexagons a heptagon, a regular seven-sided polygon, with the same side lengths as the hexagons. Notice that this surface also is not flat. What are the differences between these two non-flat surfaces?

Notice that the surface with constant positive curvature will close in on itself, but the surface with negative constant curvature will extend out indefinitely. Another way to think about this is looking at what happens when you "flatten" a surface so that it lies in the plane. For constant positive curvature, the surface covers less area than the plane. For constant negative curvature, the surface covers more area than the plane.

Flattening negative curvature.

On the surface of a banana,
there are both positive and negative curvatures.

Why are we talking about surfaces with *constant* curvature? In order to talk about a surface having a *geometry* (spherical, Euclidean, or hyperbolic), we need it to be "the same" everywhere. There are other surfaces with both positive and negative curvatures that are not constant (for example, the surface of a banana or a pear)—we will talk about some of them later.

Hyperbolic Geometry

Gauss did not know of any surfaces with constant negative curvature, but he realized that such a surface would have interesting geometric properties. In the meantime, a new non-Euclidean geometry, called hyperbolic geometry, was independently discovered (in the late 1820s) by Janos Bolyai (1802–1860) and N. I. Lobachevsky (1792–1856), but it took over three decades before their publications gained deserved recognition. They described this new ge-

ometry theoretically (in terms of axioms), but Riemann (in 1854) described how this *hyperbolic geometry would be the intrinsic geometry of a surface with constant negative curvature that extended indefinitely in all directions.* Mathematicians searched for such a surface. This is one of many occasions when many mathematicians were looking at the same thing but from many different viewpoints until somebody realized, "This is it!"

In 1866, Eugenio Beltrami (1835–1900) showed that the surface of a pseudosphere (see the picture below) has constant negative curvature and thus locally has hyperbolic geometry. Unfortunately, the pseudosphere cannot be extended indefinitely. We will talk about the pseudosphere more in Chapter 6.

Beltrami's pseudosphere.

"Models" (Maps) of Hyperbolic Geometry

Mathematicians offered, instead of using actual surfaces, several so-called "models" of hyperbolic planes; those "models" are actually maps of the hyperbolic plane into the Euclidean plane. It is similar to how we can study the geography of our "spherical" planet by looking at a flat map. Maps from a sphere onto a plane either distort areas or distort distances.

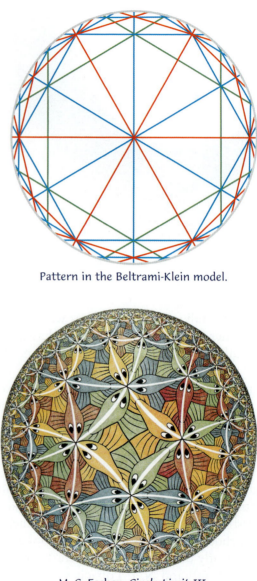

Pattern in the Beltrami-Klein model.

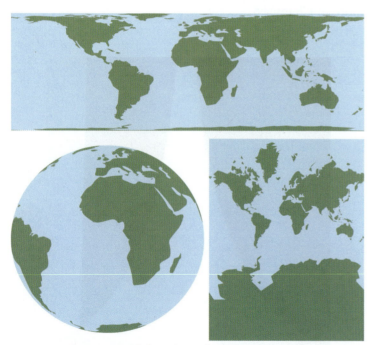

Three maps of the spherical earth onto a plane:
each distorts in a different way
(images by Chaim Goodman-Strauss).

M. C. Escher, *Circle Limit III*.
©2008 The M. C. Escher Company - Holland.
All rights reserved. www.mcescher.com

One of the first such models of the hyperbolic plane was the projective disk model or Beltrami-Klein model—Beltrami described it in 1868, and Felix Klein (1849–1925) fully developed it in 1871.

Henri Poincaré (1854–1912) suggested two models of hyperbolic geometry: the upper-half-plane model and the disk model. These models of the hyperbolic plane inspired Dutch artist M. C. Escher (1898–1972), who made several woodcuts using Poincaré's idea.

M. C. Escher, *Regular Division of the Plane VI*.

Negative Curvature in Nature

You should not think that the hyperbolic plane is only something that mathematicians or artists can come up with. You can find examples of (approximate) constant negative curvature surfaces in nature as well.

Negative curvature in lettuce.

Negative curvature in a sea slug.

Some parts of corals have negative curvature because that increases the surface area available for the coral to absorb nutrients.

Negative curvature in a holly leaf.

Negative curvature in blossoms.

Negative curvature in a cactus.

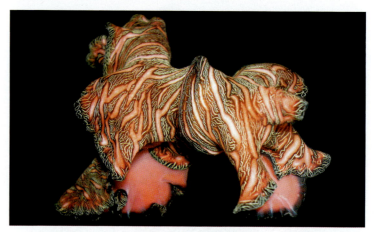

Nudibranchs are soft-bodied, shell-less sea mollusks that are known for their extraordinary colors and forms. Some of them have negative curvature (photo by Garry Cobb).

These surfaces from nature have (approximate) constant negative curvature, but, like the pseudosphere, they are not what mathematicians were looking for because they do not extend indefinitely. You can see in the pictures that the negative curvature is mostly evident around the edges.

The Search for a Complete Hyperbolic Surface

Since Descartes' and Fermat's invention of analytic geometry in the seventeenth century, mathematicians had become accustomed to describing surfaces using equations. However, in 1901, David Hilbert (1862–1943) proved that it is not possible to have an equation describe a surface in 3-space that has constant negative curvature and that is extended indefinitely in all directions (mathematicians call this "complete").[2] This theorem was improved by Holmgren,[3] who showed that, given a (local) smooth embedding of the hyperbolic plane in three-dimensional space, the embedding cannot be extended isometrically and smoothly beyond a finite distance d. Unfortunately, d depends on the local embedding, and there is not a uniform bound for the size of the "largest" piece of the hyperbolic plane that can be isometrically embedded in 3-space. Hilbert's theorem was improved by Amsler,[4] who showed that every sufficiently smooth immersion of the hyperbolic plane into 3-space has a singular "edge," i.e., a one-dimensional submanifold beyond which the embedding is no longer smooth.

Many mathematicians decided that Hilbert's theorem must mean that it is not possible to have a complete surface of a hyperbolic plane (surface with constant negative curvature) in our three-dimensional Euclidean space.

However, in 1954, the Dutch mathematician Nicolaas Kuiper (1920–1994) showed that there is a possibility of such a surface but he could not describe how to construct it. In 1956, he came up with a more general result that was analytically extended by John Nash in 1966. John Nash is better known to wide audiences due to Sylvia Nasar's novel *A Beautiful Mind* (and the movie that followed with the same title). However, Nash was interested in differential geometry in addition to the mathematics of game theory and partial differential equations, his applications of which won him the Nobel Prize in economics. This "Nash-Kuiper Theorem" is a very general result[5] that among many other things implies that there is a complete surface with hyperbolic geometry (constant negative curvature) in some (possibly high-dimensional) Euclidean space.

Soon after that, in the 1970s, William Thurston came up with the idea of using paper annuli to describe a hyperbolic surface in 3-space without any equations. For details of Thurston's construction and instructions on how you can make one, see the appendix.

This paper annuli idea was the one that inspired me to crochet a hyperbolic plane.

Hyperbolic plane made with paper annuli.

Crocheting a Hyperbolic Plane

Many times I have been asked the question of how I came up with the idea to crochet a hyperbolic plane. I knew that to crochet ruffles one must put extra stitches into each row. Then, studying the annuli, I realized that it is necessary to increase the number of stitches from one row to the next by the same ratio. I was thinking of how to visualize that and remembered an old example of how fast gossip can spread. Let us say that I heard some "news" I am not very sure about; therefore, I just tell that to only one neighbor and ask the neighbor not to tell anybody else. Now there are two of us who know the "news." On the next day, I tell just one more person and the neighbor does the same. Now there are four of us who know "the news." The following day, each of us tells just one more person, now there are eight of us. When, on the next day, the eight of us tell just one more person, then the total number of "informed persons" is already 16, and with each day the number of "informed persons" grows faster and faster. The number of persons I have talked to grows linearly, but because everybody else does the same, the total number grows faster and faster at a regular rate. (This kind of growth is called *exponential growth*. We will talk about exponential growth more after you have crocheted your model.) This is better seen if you look at the picture.

First, I saw this picture as a mathematical graph. Suddenly, I saw that this graph can be a crochet pattern where each line segment denotes a stitch. And there it was—the pattern for the hyperbolic plane. All that was left was to try it. First, I tried to knit since I am an avid knitter. But the number of stitches on the needles soon became unmanageable, and I was afraid that as soon as I accidentally lost a stitch, the whole work would unravel. So, I decided to

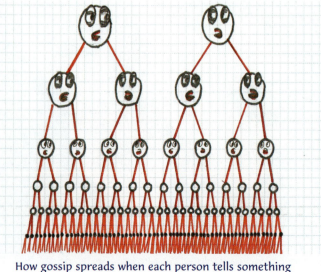

How gossip spreads when each person tells something only to one more person.

My first model of the hyperbolic plane made in 1997.

crochet because it gives more freedom in space and I had to deal only with one stitch at a time. For my first crocheted hyperbolic plane, I chose to increase not in every stitch but in every other stitch. I started with a chain 10 inches long. Ruffles appeared very quickly. After the first couple of rows, it took me longer and longer to complete the next row. The thirteenth row was 208 inches long, and I gave up on continuing this model. It was clear to me that I should start with a shorter initial chain and change the rate of increase.

In order to make the crocheted hyperbolic plane, you need just a few very basic crocheting skills. All you need to know is how to make a chain (to start) and how to single crochet. See the following figures (A, B, and C) for a picture of these stitches, which will be described further in the next paragraph. If this is your first crochet project, then I would suggest you practice a little bit until you can make your stitches evenly tight. The trick to making a hyperbolic plane is to make the crochet tight and even, keeping a constant ratio of increased stitches.

To make your handicraft tight, I suggest that you choose a smaller number hook than the one that is suggested on the label on a skein of your yarn.

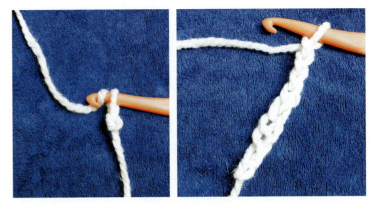

B. Crocheting a chain.

C. Starting the first row with a single crochet stitch.

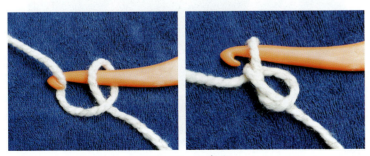

A. Starting a chain.

You should choose a yarn that will not stretch a lot. Every yarn will stretch a little, but you need one that will keep its shape. Acrylic yarns in craft stores work fine, and they come in all different colors.

That's it! Now you are ready to start crocheting your hyperbolic plane:

1. Make your beginning chain stitches. About 20 chain stitches for the beginning will be enough. (Topologists may recognize that these are the stitches in the Fox-Artin wild arc! R. H. Fox and E. Artin first described it in 1948, but for crocheters it has always been there.)[6]

2. For the first stitch in each row, insert the hook into the second chain from the hook. Take the yarn over and pull through the chain, leaving 2 loops on the hook. Take the yarn over again and pull through both loops. One single crochet stitch has been completed.

3. For the next N stitches, proceed exactly like the first stitch except insert the hook into the next chain (instead of the second one). You get to choose N; a good choice for your first hyperbolic plane is $N = 5$.

4. For the $(N + 1)$st stitch, proceed as before except insert the hook into the same loop as the Nth stitch.

5. Repeat Steps 3 and 4 until you reach the end of the row.

6. At the end of the row, before going to the next row, do one extra chain stitch.

7. When you have the model as big as you want, you can stop by just pulling the yarn through the last loop.

Be sure to crochet fairly tightly and evenly. That's all you need in terms of crochet basics. Now you can go ahead and make your own hyperbolic plane. You have to increase (by the above procedure) the number of stitches from one row to the next in a constant ratio, N to $N + 1$; this ratio $N/(N + 1)$ determines the radius of the hyperbolic plane. (See Chapter 2 and the appendix). You can experiment with different ratios *but* not in the same model. You will get a hyperbolic plane *only* if you increase the number of stitches in the same ratio all the time. For example, if you start with ratio 5:6, it means you crochet five single stitches, but when you crochet the sixth one, you insert the crochet hook twice in the sixth loop.

Crocheted hyperbolic plane with ratio 12:13.

Crocheted hyperbolic plane with ratio 5:6.

Crocheted hyperbolic plane with ratio 3:4.

Crocheting will take some time, but it is worth it because later you can work with this model without worrying about destroying it. You can see some examples of finished planes in the preceding pictures.

Exponential Growth

One thing you probably already noticed is that as you crochet, the number of stitches in each row grows very fast. As we mentioned before, in mathematics, this faster and faster increase is called exponential growth or geometric growth. It means that the number of stitches in each row grows proportionally to its size. This implies that the more rows there are, the faster the number of stitches in a row grows, but it also implies that the ratio of the number of stitches in one row to the number of stitches in the next row is governed by a strict law—direct proportion. Remember that in the crochet pattern for a hyperbolic plane, it was stressed that you should keep the same ratio for your entire project. You should also crochet another hyperbolic plane with a different ratio and notice that the higher the ratio, the faster the number of stitches grows or the other way around—for smaller ratios, it takes longer to notice that your model is not flat but has ruffles.

One of the easiest ways to experience exponential growth is by folding a sheet of paper. Take an ordinary sheet of paper and fold it in half. Then fold it in half again. Before continuing this folding, can you guess how many times you will be able to fold this sheet? Now check your guess. Was it correct? If it would be possible to continue folding this sheet, after how many folds will the thickness be taller than the height of your house?[7]

Chapter 2

What Can You Learn
from Your Model?

A friend of the famous Swiss artist Max Bill (1908–1994) asked the artist to make a sculpture that could be hung over his fireplace. Bill started to look for ideas, and he thought that he had invented a completely new shape simply by twisting a strip of paper. To his great surprise, people started to congratulate him on a fresh and original interpretation of the ancient Egyptian symbol for infinity and the Möbius band. Bill confessed: "I had never heard of either of them. My mathematical knowledge had never gone beyond routine architectural calculations, and I had no great interest in mathematics."[1] After this experience, over the years Bill produced many other mathematical surfaces.

What does this story about Bill's sculpture tell us? You can create new shapes without really knowing the mathematics behind them. If you do not have a strong mathematical background but you have been doing crochet, you have perhaps already recognized that making the hyperbolic plane is in some ways reproducing a common beginner's mistake in crocheting a hat: if you add too many stitches, instead of being nice and round (by now you know that this would demonstrate positive curvature), the hat develops ruffles (which have negative curvature). You probably have already crocheted surfaces with negative curvature; you just did not know that mathematicians have this name for them! Once you have made your own hyperbolic plane model, I am sure you will want to find out what else you can learn from it.

Max Bill's *Endless Ribbon* (photo by Nathaniel Friedman).

What Is Straight on a Hyperbolic Plane?

> … to ask what geometry [Euclidean or non-Euclidean] it is proper to adopt is to ask, to what line it is proper to give the name straight?
>
> –Henri Poincaré[2]

How can you produce a straight line on a piece of paper? I'll take a pencil and a straightedge and will draw it, you can say. Yes, but this technique will not help to get a straight line on a hyperbolic model. Another way is just to fold a piece of paper, making a straight crease, as is done in origami (the ancient Japanese art of paper folding).[3] Geometric constructions done by folding paper were well known in medieval India.[4]

The same folding can be done with your crocheted hyperbolic plane: you fold it and the crease will show you a straight line on your hyperbolic plane.

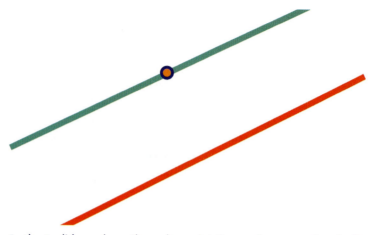

Constructing a straight line by folding paper.

In the Euclidean plane, through a point (orange) not on a line (red), you can draw only one line (green) parallel to the red line.

Parallel Lines

If you draw a straight line on a flat piece of paper and choose a point not on this line, then you can draw only one straight line through this point that will be parallel to the original straight line (i.e., that will never intersect the original line).

On a hyperbolic plane, parallel lines behave differently. In fact, the rule for parallel lines on the hyperbolic plane is that for a line and a point not on that line, *more than one* line can be drawn through the point that never intersects the first line. We can find two types of such lines on a hyperbolic plane.

One type is the lines that are perpendicular to the crocheted rows in our models: they diverge in one direction but come closer and closer in the other direction without intersecting (mathematicians say they are *asymptotic*).

Folding (without stretching) a hyperbolic plane produces a straight line on it.

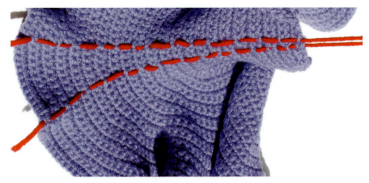

Asymptotic straight lines in the hyperbolic plane:
they become closer and closer but never intersect.

Two nonintersecting lines in the hyperbolic plane
that diverge in two directions.

The other type is the lines that are closest to each other only in one place on the plane but then diverge in both directions.

Now let us look at two nonintersecting lines on your hyperbolic plane model. Choose a point on one of the lines. Try folding your model in a way that you get another straight line through this point that will not intersect the other line. In how many ways can you do it? In the picture below you can see three such lines marked on a model, but of course, you can find as many as you like.

This is the first diagram I made for myself after I made a second crocheted hyperbolic plane model that was large enough to make diagrams on it.

When I was studying non-Euclidean geometry as a student, the professor talked about Lobachevsky imagining parallel lines between stars and those lines being curved, therefore converging and then diverging again. I remember being very confused—how can lines converge and then diverge and still be straight? I passed my exam and got credit for that class, but somewhere inside me confusion remained for many years—until I made a model and saw what is happening in the hyperbolic plane. Lobachevsky called his geometry "imaginary" because it was so greatly in contrast with common sense; so it remained for me also "imaginary," until I could experience straightness in the hyperbolic plane in a tactile way.

Lines through the same point and not intersecting another line
in the hyperbolic plane.

In the next picture you can see two yellow lines that come close to each other but then diverge. Both of these lines are perpendicular to the same (red) line. You can construct a perpendicular to a given line on the hyperbolic plane the same way you would when folding paper: fold the model to create the first line, then fold again so that the first crease lies on itself. The second fold line is perpendicular to the first. You can mark these lines by weaving yarn of a different color on your model, as was done in these pictures.

Draw two parallel lines on a piece of paper: how many common perpendiculars can you construct for these two lines? If you do it in different places, then, of course, you can draw as many as you want. What is common to these perpendiculars? If you measure them between the two parallel lines, then all of them are the same length. This property is sometimes used in practice to construct straight lines on fields. Does the same property hold on a hyperbolic plane? A common perpendicular means that you can fold along it and both lines will lie on themselves.

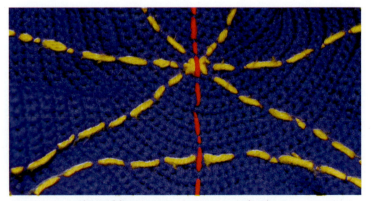

The red line is a common perpendicular to only two of these yellow lines.

Angles of a Triangle

The next property to explore is the sum of interior angles of a triangle. First draw a triangle on a piece of paper, mark its angles as α, β, γ, and then cut the triangle out. Now tear off its angles and align them next to each other so that the nonripped corners meet (and the pieces don't overlap). You can see that the sides of the outside angles form a straight line. Equivalently, we can say that the sum of three interior angles of the triangle is 180 degrees. In the flat Euclidean plane, no matter what triangle you choose, the sum will always be the same: 180 degrees.

Let us look now at the surface of a sphere. Straight lines on a sphere are great circles. You can check this by experimenting with a tennis ball and a rubber band—only when the rubber band lies on a great circle will it stay on the tennis ball. You can see this more easily by looking at the globe, where longitudes and the equator are straight lines because each of them divides the globe exactly in two halves. Latitudes (except the equator) are not straight lines but just circles on a sphere. The triangle ABC on the globe in the photo has two, sides each of which is a segment of a longitude from the North Pole, A, to a point on the equator. The third side is the segment joining these points, B and C, on the equator. The longitudes intersect the equator at right angles. Thus, this triangle has two right angles at its base (segment on the equator) plus the angle between the meridians at the North Pole. For this spherical triangle, the sum of the interior angles is greater than 180 degrees.

Is there another possible surface on which the sum of the angles of a triangle can actually be less than 180 degrees? Gauss was thinking about this possibility when in 1824 he wrote to his friend:

There is no doubt that it can be rigorously established that the sum of the angles of a rectilinear triangle can-

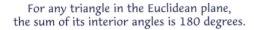

For any triangle in the Euclidean plane,
the sum of its interior angles is 180 degrees.

Spherical triangle, ABC, on a globe
has an interior angle sum greater than 180 degrees.

not exceed 180°. But it is otherwise with the statement that the sum of the angles cannot be less than 180°; this is the real Gordian knot, the rocks which cause the wreck of all... I have been occupied with the problem over thirty years and I doubt if anyone has given it more serious attention, though I have never published anything concerning it.[5]

You can read more about this in David E. Rowe's article "Euclidean Geometry and Physical Space."[6]

Now try to construct different triangles on your hyperbolic plane model: pick three points and mark a straight line between each pair of those points. What do you notice? Try to make as big a triangle on your model as possible. What is happening to the sum of the angles of this triangle compared to smaller triangles? Can you make as big a triangle as you want? I really encourage you to try it.

The largest possible triangle on your hyperbolic plane is called an *ideal triangle*. You can notice that the interior sum of the angles of this triangle is approaching zero. If you folded one ideal triangle on your model, you can try to make another one on the same hyperbolic plane. You will notice that you can try as hard as you want, but these triangles will be close to the same size.

Different Radii

If you have another model of a hyperbolic plane that is made with a different ratio of increases, fold an ideal triangle on this plane also. You will notice that the ideal triangles on the same plane are the same size, but on the two hyperbolic planes made with different ratios, they will be different. Why are they different? What does this observation tell us about the hyperbolic plane?

First, let's define the radius of a hyperbolic plane, which can vary, for example, depending on how the model is crocheted. When you look at your crocheted model of the hyperbolic plane, you can see that the rows you stitched are not straight but that they bend—they are like arcs with the same radius, and we can call them *annuli*. The radius of the hyperbolic plane is the radius of these annuli.[7] You can approximately measure the radius of a hyperbolic plane by

An ideal triangle in the hyperbolic plane:
the sum of its angles approaches 0 degrees.

The radius of the hyperbolic plane is the same as the radius
of a row of stitches (if the arc continued into a full circle).

How to determine the radius of the hyperbolic plane: here, the radius is about 2 inches.

Many familiar objects are spheres with different radii.

placing it on a flat surface. Along where the plane touches the surface, a little three-dimensional arc will form. Put a thread around it, forming a full circle, then measure the diameter of the circle and divide that value in half. That will be the radius of your hyperbolic plane.

In many textbooks on hyperbolic geometry, it is stated that the area of an ideal triangle in the hyperbolic plane is π. This is only true with the assumption that the radius of the hyperbolic plane equals 1. But to talk about all hyperbolic planes having the same radius would be the same as assuming that all spheres have the same radius! You can measure the radius of your hyperbolic plane as shown in the pictures.

With spheres, we can notice that the smaller the sphere, the more it is curved. If the radius of the sphere increases, it becomes less curved. The larger the radius of the surface of a sphere, the flatter it feels. When we stand on the surface of the Earth, we do not notice it to be curved because the radius of the Earth is too large for us to experience the curvature of the Earth. So we can say that if the radius of the sphere is infinitely large, then curvature approaches zero. Or, we can equally define the Euclidean plane to be a sphere with infinitely large radius. Recall that if a sphere has radius R, then its curvature is defined by Gauss to be $1/R^2$. The curvature of the hyperbolic plane with radius R is $-1/R^2$. Notice that, in this case also, it is true that the larger the radius of the hyperbolic plane, the flatter it becomes, and if the radius becomes infinitely large, then the hyperbolic plane approaches the Euclidean plane.

Hyperbolic planes with different radii.

the sphere will be less than that of the circle you drew on the paper. The circle on the sphere with the same intrinsic radius encloses less surface. Remember the picture from Chapter 1, "Flattening positive curvature."

On the hyperbolic plane, the opposite will happen: the circumference of the circle will be larger than the circumference of the circle with the same radius on the paper and also the area enclosed by this circle will be larger than the area enclosed by the circle on the paper. Remember the picture from Chapter 1, "Flattening negative curvature."

There is another way we can experience curvature. Take a piece of paper, a ball, and your hyperbolic plane. Now, use a fixed length of string—pinning down one end, almost like a compass—to draw a circle on the paper, then on the surface of the ball, and then on your hyperbolic plane.[8]

One of the most remembered results from school geometry, known since Babylonian times (more than 4000 years ago), is that, in the plane, the ratio between the diameter of a circle, d, and its circumference (perimeter of the circle), C, is always constant, and this ratio is the famous constant π: $C = \pi d$. The area of the circle with radius r in the plane is πr^2.

On a sphere, a positively curved surface, this formula will not work: the circumference and area of the circle on

A circle on a sphere with a radius of 7 cm.

A circle with the same radius (7 cm) on the hyperbolic plane.

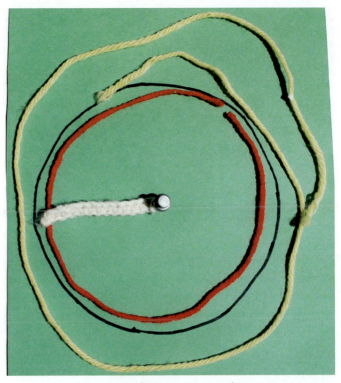

Circumferences compared.

The last picture shows comparisons of the circumferences from the three surfaces. The black circle is the circle with a radius of 7 cm in the Euclidean plane. The red thread indicates the circumference of the circle with the same radius on the sphere, and the yellow thread is the circumference of the circle with the same radius on the hyperbolic plane.

As you can see from the explorations in this chapter, you can learn things just from your own experiences. That is the way science starts—from observations, experience, collecting data, and then generalizing it. The next chapter will talk about different human experiences that led to the development of geometry.

Chapter 3

Four Strands in the History of Geometry

The ways in which different ideas have become abstract geometric concepts depend on the ways in which they have been explored. H. Graham Flegg wrote:

> New branches of mathematics come into being, not because they are created overnight out of nothing by some individual genius, but because the soil has been prepared over the previous decades (or even centuries) and because some internal or external stress (or perhaps a combination of both) provides the appropriate impetus and motivation at the crucial point in time. More often than not, it is the case that several minds produce independently and almost simultaneously the germs of what subsequently develops into a new theatre of mathematical investigation. For this reason it is usually ill-advised to point to any one man as being the founder or inventor of any particular branch of mathematics.[1]

This chapter gives suggestions of how certain geometric ideas might have come into existence—we really do not have the ability to go back in time and trace the road of knowledge again. There are many unanswered questions related to the origins of geometry.

However, it is helpful to think of the main aspects of geometry today as emerging from four strands of early human activity that seem to have occurred in most cultures:

- art/pattern strand,
- building/structures strand,
- navigation/stargazing strand, and
- motion/machines strand.

These strands developed more or less independently into varying studies and practices that eventually, from the nineteenth century on, were woven into what is now called *geometry*. Geometry today means much more than the plane and solid Euclidean geometry familiar to us since high school math classes.

Let us look at the different experiences that led to development of different geometric ideas, and then we will see how the crocheted hyperbolic plane connects with these various geometric experiences.

Art/Pattern Strand

> Man's earliest ambition is to create. To this feeling must be ascribed the tattooing of the human face and body, resorted to by the savage to increase the expression by which he seeks to strike terror on his enemies or rivals, or to create what appears to him a new beauty.
>
> —Owen Jones[2]

The English architect and designer Owen Jones (1809–1874) argued that humans could first learn about the symmetries of their own bodies, and those of animals, from reflections in the water and other observations in the surrounding world.

The formation of patterns by the equal division of similar lines, as in weaving, would give to early humans other notions of symmetry and repeating patterns. To produce decorations for their weaving, pottery, and other objects, early artists experimented with many symmetries and repeating patterns. The simplest geometric elements, such as line segments and triangles, would be joined by curvilinear figures for use in ornaments, for example, to decorate tools and weapons. Ancient drawings, paintings, sculptures, and ornaments are less than 100,000 years old—some found in Africa, Australia, the Middle East, and Europe. One of the oldest ornaments is a disc ornament in the British Museum, approximately 13,000 years old. It was found in Montastruc, France.

Disc ornament, approx. 13,000 years old.

Over time, patterns possibly first used to mark the property of a certain family or tribe became culturally charged—they were not solely geometric symbols arranged in some order, but they also reflected the environment where a particular culture was created. For example, Incan patterns reflect the angularity of the mountains where they lived. The strip decorations in Maori patterns reflect the waves of the sea surrounding New Zealand.

Geometric patterns on ancient pottery
in the National Museum of Scotland, Edinburgh.

Maori pattern.

Geometric designs with repeating patterns are still used in contemporary pottery in the Philippines.

The Inca civilization did not have writing but did have different ways of preserving and conveying important information through collections of various, mostly geometric, symbols, for example, *tucapu*. The tucapu were used to make distinctions of rank and political organization.

A basketry-covered wooden bucket with geometric patterns from the Pacific islands was collected by Captain Cook on his second voyage (1773–1774). The wooden container is encased in plaited coconut fiber, some of which is dyed black. The triangular patterns are created by contrasting dyed and undyed fiber. These are accentuated with small sea shell and coconut shell beads. In the Tongan Islands, objects have always reflected the social status of their owner, thus both the designs, and the right to use this kind of container, would have been dependent on the owner's position in society.

Astray jars from the Philippines (photo by Imes Chiu).

Inca tucapu design.

Basketry-covered wooden container from the Tongan Islands
©2008 The Trustees of the British Museum.

Many modern Navajo designs are elaborations on patterns found in ancient Navajo baskets and petroglyphs from the Navajo Nation. Navajo weavers say that any design woven by a Navajo weaver within the four sacred mountains of the Navajo Nation is sacred and that the plants, animals, rocks, mountains, and other geographical elements of the region, as well as the sheep wool, dye plants, and pinion pine used to build the weaving loom, were given to the Navajo by their Holy People.

Stone Age artists carved interesting geometric designs such as spirals on large stones in their passage graves. Such passage graves have been found in Newgrange in Ireland and to the north of Scotland in the Orkney Islands.

Spirals and concentric circles in the Lintel Stone, ca. 3000 BC, from the Orkney Islands (National Museum of Scotland).

Traditional Navajo design.

Neolithic ball with spirals in the National Museum of Scotland.

The significance of these spirals is not entirely clear but is very likely associated with the seasonal decline and subsequent rise of the sun, and its very regular periodic motion may be a symbol for the life cycle.

Nautilus shell.

Ferns start out as spirals.

A fossil, approximately 450,000 years old, in the British Museum.

Spirals possibly were copied as ornaments from nature, where they can be seen in some ferns and in the shape of a nautilus shell. Spectacular spirals can be found as fossils.

Spirals were also used to decorate the beautiful ceramic pots made by the Native American people who lived about 1000 years ago in the lost city known as Cahokia in Missouri.

Spirals were used to decorate Mycenaean Greek jars as early as the fifteenth century BC. Later, arcs and circles were used by the Greeks to decorate amphora (or storage vessels) and kraters (or two-handled jars for mixing wine with water) in the tenth century BC, known as the Protogeometric Period. Then, more elaborate geometric patterns, such as meanders, zigzags, and triangles, as well as animal and human figures, were used to decorate kraters in the ninth and eighth centuries BC in the full Geometric Period.

We can view many geometric symbols as a way to convey ancient cosmological knowledge. Later, these symbols became traditional ornaments. For example, see the table with cosmological symbols of the ancient Balts. Those signs are very similar to some signs used in Sanskrit and also by Native Americans.

They also appear in a sash that was used as a belt in some Latvian national costumes, and those same signs are an essential part of designs in Latvian mittens and Turkish socks. Ancient Celtic symbols are also mostly geometric.

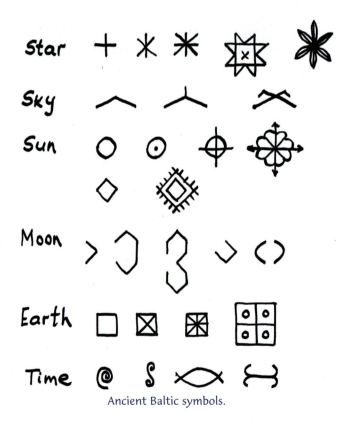

Ancient Baltic symbols.

Latvian sash "Lielvardes josta."

Celtic art spans almost 17 centuries from the early Iron Age to the Anglo-Norman invasion of Ireland in the twelfth century. Celtic art is characterized by a simple balance of abstract patterns, refusing all attempts to mimic the natural or animal environment.[3]

The sona illustrate proverbs, fables, games, riddles, and stories about animals. The storyteller starts by making a series of dots, evenly spaced, in a rectangle. Each drawing has a set number of rows and columns. The drawing itself is made up of a continuous line drawn around the dots. You cannot connect the dots or trace the same line segment twice.[4] The storyteller must draw and talk at the same time. This tradition has played an important role in the passage of knowledge from one generation to the next.[5]

Ancient Celtic symbols in the National Museum of Scotland.

The Tchokwe people of northeast Angola are famous for their decorative arts, including beautiful woven mats and baskets, pottery, and wood sculptures, and for the striking geometric designs they use to decorate the walls of their homes. Beaded masks and decorated basketry have interesting traditions elsewhere in Africa, as well.

The Tchokwe people have the curious ancient tradition of using drawings in sand to illustrate their stories. These drawings are called the *sona*.

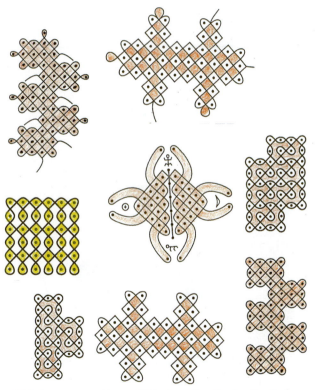

Tchokwe sona patterns (drawing after Paulus Gerdes).

For ancient artisans, decorating different shapes was also the experience of geometry on different surfaces. For example, we can see curvilinear triangles on Chinese pottery from the Neolithic Period.

Neolithic people in Scotland created interesting spherical shapes with ornaments.

Earthware form China, ca. 3500-2800 BC
(H. Johnson Museum of Art, Cornell University).

Neolithic stone balls in the National Museum of Scotland.

In antiquity, mosaics were designed to decorate architectural surfaces. Most characteristic are the mosaics that are made up of *tesserae*—pieces of stone, glass, or terracotta cut more or less to the shape of a cube. This was the form of mosaic chosen for the rendering of the geometric patterns, vegetal motifs, and figure compositions used in pavements from the fourth century BC to the early Christian period. Ancient mosaics can be divided broadly into two main categories: those that decorated floors and those that decorated walls or vaults. In pavements, by far the most common form of treatment was geometric. Motifs could range from simple checkerboards to complex compositions involving lozenges, octagons, hexagons, squares set diagonally, L-shapes, and roundels or other curvilinear shapes; arrangements could be continuous or they could be centralized.

Mosaics can be seen in many ancient cultures. The earliest examples of mosaics are found in Sumerian architecture of the third millennium BC as decorations of columns. Pebble pavements with random patterns had appeared as early as the eighth century BC in Gordium, capital of the native kingdom of Phyrgia in Asia Minor. But the first ordered patterns, and the first representation of figures and animals in mosaic, appeared around the late fifth and early fourth centuries BC in cities of the Ancient Greek world.

The richest, largest, and most complex collection of late Roman mosaics in the world is the Villa Romana del Casale, at Piazza Armerina, Sicily.[6] The mosaics were probably made by African artists in the early fourth century CE. The North African provinces were at the economic and artistic forefront in the fourth century, and polychrome mosaics were one of the specialties of the North African artists. Very similar mosaics have been found in Carthage and other places in North Africa.

Fourth-century Roman mosaics from Piazzo Armerina, Sicily.

The Byzantine cathedral (1094) in Venice on Piazza St. Marco is very famous for its mosaics, but mosaics were used also in other palaces. Among different patterns, I was surprised to notice one that shows the first steps of the so-called *Sierpinski triangle*. This is an interesting geometric pattern that can be constructed connecting midpoints of the sides of an equilateral triangle. Polish mathematician Wacław Sierpiński (1882–1969) described this in 1915, and it is one of the first fractal constructions. Mosaic makers in the eleventh century could not possibly know modern mathematics!

Sierpinski triangle pattern in the Byzantine cathedral, Venice.

Mosaics from Piazza St. Marco, Venice.

Islamic art is strongly based on various geometric figures such as equilateral triangles, squares, and many different regular polygons with sides ranging in number from 5 to 24. Geometric patterns can be found on diverse materials: tiles, bricks, wood, brass, paper, plaster, and glass. They were used on carpets, windows, doors, screens, railings, bowls, and furniture. Beautiful patterns appear in Persian miniatures. Symmetric Islamic patterns can be seen in three different applications. One way to create a symmetric pattern is the so-called calligraphic approach that would often create symmetric geometric forms by molding the Arabic lettering for words such as "Allah" or "Mohammed" or for a short verse from the Koran. Another perfected pattern in Islamic ornaments is the Arabesque. These patterns are characterized by their intertwining spiral forms that produce repeated stylized leaves and flowers. The third and the largest class of Islamic patterns are polygon-based space-filling patterns that also sometimes use regions bounded by circular arcs. The most frequent shapes in these geometric patterns are stars (most often with 5, 6, 8, 10, 12, or 16 rays) and rosette shapes. Some tessellations are based also on other numbers, particularly on the multiples of eight up to 96. Splendid examples of Islamic tessellations can be found in Spain, for example, in the Alhambra in Granada, and elsewhere in the Muslim world.[7]

M. C. Escher first visited the Alhambra in 1923. He was amazed by the Moorish tiles and their intricate geometric patterns. In Escher's words, "The Moors were masters of filling a surface with geometric figures." He admired the tile work and the beautiful tessellations and was inspired to work with tessellations himself.

Islamic pattern on a wooden panel in the National Museum of Scotland.

Pazyryk carpet (fourth or fifth century BC).

In 1949, a Russian archeological expedition in the Altai Mountains found a royal burial site in which a preserved carpet known as the Pazyryk carpet was found. It was used as a saddle cover. This rug dates from the fourth or fifth century BC and is the earliest known surviving example of a hand-knotted carpet.

Later, the study of symmetries of patterns led to such mathematical concepts as tilings, group theory, crystallography, and finite geometries.

The early artists also explored various methods of representing physical objects and living things. In some Roman mosaics, we can already notice attempts to represent three-dimensional objects.

For tent dwellers, the most natural form of furniture is carpet. It is therefore not surprising that the art of making carpets has been very popular throughout the Middle East and the Caucasus region. The nomadic tribes of central Asia, Persia, and Afghanistan have been producing carpets and rugs as floor coverings, prayer mats, and tent decorations for thousands of years. Over that time, patterns of these carpets became more and more elaborate since carpets also became symbols of power, privilege, and wealth. Later, ornaments on Islamic buildings could be considered as reflecting the tradition of carpets.

Mosaic from Piazzo Armerina, Sicily.

The mathematical development of the geometry of perspective originated in the works of Renaissance sculptors, painters, and architects, like Filippo Brunelleschi (1377–1446), Battista Alberti (1404–1472), Paolo Uccello (1397–1475), Piero della Francesca (c. 1412–1492), Leonardo da Vinci (1452–1519), and Raphael (1483–1520).

Piero della Francesca wrote a book on perspective (*De Prospectica pingendi*, c. 1474) and a book on the five regular or Platonic solids and five of thirteen Archimedean solids that are truncated Platonic solids (*Libbellus de quinque corporibus regularibus*, c. 1480).

German artist Albrecht Dürer (1471–1528) was very interested in mathematics, like the Italian artists before him, and wrote a book on geometry (*Underweysung der Messung*, c. 1525), treating the subject of measurement with only compass and straightedge as well as other geometric topics. Dürer also wrote a book in four parts on human proportion from a truly geometric point of view (*Vier Bücher von menschlicher Proportion*, 1523).

Dutch artist Jan Vermeer (1632–1675) painted many pictures that showed his good understanding of perspective. These artistic explorations in the Renaissance led to the study of perspective and then later to projective geometry and descriptive geometry in mathematics.

Geometry had an impact on modern art. Interplay between art and geometry happened in the late nineteenth century and the early twentieth century, when ideas about non-Euclidean space and higher-dimensional spaces gradually began to appear in nonmathematical literature. For artists, the possibility of non-Euclidean space meant that the great achievement of the Renaissance—linear perspective—would be invalid, and that became an inspiration for artistic experiments. Ultimately, the idea of higher-dimensional spaces became far more popular than the notion of curved non-Euclidean space.

The Cubists saw the relationship of non-Euclidean geometries to Euclidean geometry as parallel to their own situation in the history of art.[8]

In the last two centuries, this art/pattern strand has led to security codes, digital image compression, computer-aided graphics, the study of computer vision in robotics, and computer-generated movies.

Building/Structures Strand

For human life it was essential to make tools. Ancient stone tools demonstrate nice bilateral symmetry.

Hand ax, found in France, about 350,000 years old
(British Museum, London).

Hand ax, found in Britain, about 6000 years old (British Museum).

Fire altar design idea from *Sulbasutram.*

As humans built shelters, altars, bridges, and other structures, they discovered ways to make circles of various radii and various polygonal/polyhedral structures. In the process, they devised systems of measurement and tools for measuring. The massive stones in Stonehenge and in Northern Scotland were assembled in circles so accurately that they have survived for thousands of years without significant movement. They testify to the mathematical understanding of the stresses and strains in the megalithic construction by the Neolithic engineers who designed it.

The ancient Indian manuscript *Sulbasutram* (2000–600 BC) was written for altar builders. At the beginning, it contains a geometry handbook with proofs of some theorems and a clear general statement of the "Pythagorean" Theorem. (A. Seidenberg argues that geometry had a ritual origin.[9]) Constructions described in *Sulbasutram* were for altars of various shapes depending on the particular ritual for which they would be used. This manuscript also described a problem of converting a rectangular shape to the square with the same area.

The problem of squaring the circle perhaps originated in building burial places—in Northern India there was an ancient tradition of burying distinguished persons in square mounds, but heretics were buried in circular mounds with the same area.[10] The problem of squaring the circle is one of the three famous construction problems in antiquity. The other two are trisecting an angle and duplicating the cube. Later, large parts of Greek geometry were built around these problems.

Geometric constructions are an important part of the ancient art of constructing a *mandala*—meaning "circle" in Sanskrit. The mandala pattern has been used in several religious practices. Best known are Tibetan sand mandalas used for spiritual practice to demonstrate the impermanence of life. Native Americans also created sand mandalas and medicine wheels. The circular Aztec calendars were both timekeeping devices and an expression of religious beliefs.

Tibetan monks spent many hours creating this mandala honoring the Dalai Lama's visit to Cornell University in 2007.

and sweathouses all employed the use of the circle and oval, each structure serving as a gentle reminder to users of their place in the larger circle of life. Village and camp centers were also laid out in a circular fashion, reminding the villagers of the four sacred points of the compass, the winds, the four seasons, and the universe in which they were full partners.[11] As Black Elk said, "… the Power of the World always works in circles, and everything tries to be round."[12]

Circular shapes were in use in the Neolithic period in Ancient China in the Liang Zhou culture, but little is known about this culture.

For the Native Americans of the Great Plains, ritual origins can be seen also in building the *tipi*—a conical-shaped tent originally built from animal skins or birch bark. The spatial geometry of the tipi was used to symbolize the universe in which the Plains people lived. The home was a place of worship as well as a place to merely eat and sleep. The floor of the tipi represented the earth—the Mother. The lodge-cover was the sky above—the Father. The poles linked mankind with the heavens, representing the four directions. They usually placed the altar to the west, where the sun sets, thus keeping the fire alive all through the night in the place where the sun disappeared.

Numerous sacred dances incorporated the circle in the form and use of the hoop and/or dance and movement around the center of the space in which many ceremonies took place. The tipi, medicine wheel, sundance structures,

Objects of unknown purpose from the Liang Zhou culture, Neolithic Period, China, ca. 2500 BC (British Museum).

All Gizah Pyramids (photo by Ricardo Liberato).

Ancient building in Uxmal, Mexico (photo by Pedro Sánchez).

The Parthenon, Athens.

The construction of such structures requires very good knowledge of angles and proportions. Some of these proportions also had religious significance. One of the most popular proportions used in ancient architecture was the so-called *golden ratio*. There is much literature[13] available about the golden ratio in architecture, nature, art, and other connections. One particularly famous use of the golden ratio is the exterior dimensions of the Parthenon in Athens.

There are many other excellent examples of ancient building skill and geometry from around the world, for example, in Mexico.

Building upon geometric knowledge from Babylonian, Egyptian, and early Greek builders and scholars, Euclid (325–265 BC) wrote his *Elements*, which became the most used mathematics textbook in the world for the next 2300 years and codified what we now call Euclidean geometry.

Using the *Elements* as a basis, in the period 300 BC to about 1000 CE, Greek and Islamic mathematicians extended Euclid's results, refined postulates, and developed the study of conic sections and geometric algebra. The first full mathematical theory following Euclid's tradition was Apollonius' (ca. 262 BC–ca. 190 BC) *Conics*.

Within Euclidean geometry, analytic geometry, vector geometry (linear algebra and affine geometry), and algebraic geometry developed later. The *Elements* also started what became known as the axiomatic method in mathematics—a method in which a few basic facts are given to be true and then other statements are proved (or disproved) based on those facts. Euclid's traditions were so strong that even Sir Isaac Newton wrote his famous *Principia Mathematica* (1687) in the language of Euclidean geometry. Eighteenth-century French mathematician, physicist, and philosopher Jean le Rond d'Alembert (1717–1783) wrote that the *Principia* is "the most extensive, the most admirable, and the happiest application of geometry to physics which has ever been made."

For the next 2000 years, mathematicians attempted to prove Euclid's Fifth (Parallel) Postulate as a theorem (based on the first four postulates);[14] these attempts culminated around 1825 with the discovery of hyperbolic geometry. This will be discussed more in Chapter 5.

Further developments with axiomatic methods in geometry led to the axiomatic theories of the real numbers and analysis and to elliptic geometries, axiomatic projective geometry, and other axiomatic geometries.

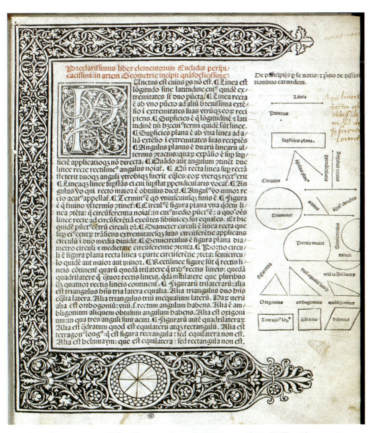

Opening page of the first printed edition of Euclid's *Elements*, 1452, Venice (courtesy of the Division of Rare and Manuscript Collections, Cornell University Library).

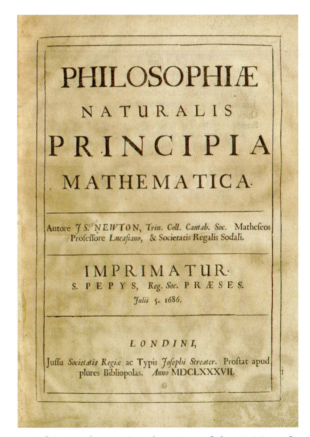

Title page of Newton's *Principia* (courtesy of the Division of Rare and Manuscript Collections, Cornell University Library).

Navigation/Stargazing Strand

Life happens in cycles. Early humans must have noticed that all these cycles, whether they are human life, animal, plant, solar, lunar, or seasonal, have different characteristics, but they have a common theme: birth, growth, death. Early humans exercising their reasoning must have noticed that they can get more control over their environment if they can predict certain things happening, like season changes. Predicting lunar and solar eclipses was central to early spiritual practices. But predicting meant being able to record, measure, and compare intervals of time, angles, and distances. Recording the passage of time was needed for humans to be ready for natural events like tides, rains, floods, growing season, and hunting season. Since time is related to the motion of heavenly bodies, it is natural that astronomy and timekeeping should have developed at the same time.

The easiest way to mark time was with shadows. In Ancient Egypt, special shadow boards were used for measuring the time taken to perform tasks or for timing the distribution of water for irrigation. Another method of measuring the apparent position of the sun was to use a shadow stick or *gnomon*. This primitive device was used to see how much daylight remains in order to set up some communal events, such as mealtimes. Sundials use a gnomon to cast shadows on a marked plate.

For political, religious, agricultural, and other purposes, ancient humans attempted to understand the movement of heavenly bodies (stars, planets, sun, and moon) in the apparently hemispherical sky. Early humans used the stars and planets as they started navigating over long distances on land and on the sea. They used this understanding to solve problems in navigation and in attempts to understand the shape of the Earth.

Nearly 4000 years ago, the Babylonians developed the sexagecimal system of angular measurement defining the full circle as 360 degrees (with 1 degree = 60 minutes and 1 minute = 60 seconds), corresponding approximately to the angular movement of the earth during one day in its orbit around the sun. The Babylonians had adopted the vernal equinox, marking spring as the start of their year, and the zodiac. The astronomical developments in Babylon were led from the temple and were interlinked with religion and several gods of the time.

Seventh century BC Babylonian map,
© The Trustees of the British Museum.

Babylon was at its zenith between 1900 and 1600 BC, but for the following thousand years, Mesopotamia was like a battlefield. Finally, in 539 BC, it fell to the Persians, who established the greatest empire then known through most of the Middle East.

The other great early civilizations, such as those in Egypt, India, and China, also conducted astronomical studies that were driven by practical, astrological, and religious motives. Early Chinese cosmology assumed the universe was a rotating sphere with fixed stars.

Aristarchus (ca. 310–230 BC) became a great scholar in Alexandria. People before him had asked questions like how far is it to the moon? To the sun? To the stars? But Aristarchus was the first one to devise geometric methods to answer them. Aristarchus, for example, noticed that in a lunar eclipse the sun, Earth, and moon are in a line, with the moon appearing full. But when the moon passes into the shadow of the Earth, the relative sizes of the bodies can be estimated from the curvature of the moon's bright disk and the curvature of the Earth's dark shadow on it. Aristarchus concluded that the Earth's diameter was three times larger than the moon's. Actually, it is nearly four times larger, but Aristarchus' result is quite reasonable for the methods available to him. One of the difficulties of obtaining more accurate results was that the Earth's diameter was unknown.

About 250 BC, a young mathematician, Eratosthenes (ca. 273–195 BC), arrived in Alexandria. He knew that the earth was round—that had been proposed already by the Pythagoreans. Eratosthenes came up with a method to estimate the radius of the earth. He had noticed that at Syenne (present-day Aswan) the sun was directly overhead at noon on the midsummer solstice. This means that a vertical pole in Syenne would cast no shadow. At the same time, a vertical pole erected in Alexandria was casting a shadow that was one fiftieth of the height of the pole. The angle involved here (about 7 degrees) represents an angle that would be the angle between poles in Alexandria and Syenne if they were to be extended to the center of the earth. Eratosthenes measured the distance between Alexandria and Syenne to be 5000 stadia, which led him to estimate the circumference of the earth to be 250,000 stadia. It is believed now that one Egyptian stadium was about 160 m. It means that Eratosthenes' result was 40,000 km (the present value is 39,940 km)—pretty remarkable!

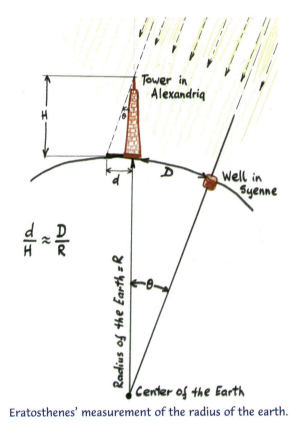

Eratosthenes' measurement of the radius of the earth.

The navigation of rivers, lakes, and oceans began before recorded history. Navigation, due to its relationship and importance to transportation, has played a leading part in the advancement of civilization. People learned early that travel by water was a convenient means of transporting their goods of trade to other lands. The people living near the Mediterranean Sea—the Sumerians, Cretans, Egyptians, Phoenicians, and Greeks—became able mariners, as did the Scandinavians in northern Europe. The early mariners did not venture very far from the coasts. Skirting the coastlines, they could identify objects on land and thereby know the positions of their ships. Usually they traveled by day and went ashore at night. They did not have nautical charts, but sometimes they found their way by a list of directions. The Romans called such a list a *periplus*. It gave details of landmarks, good anchorages, and such hazards as shoals and reefs.

The Phoenicians and Greeks were the first of the Mediterranean sailors to navigate far from land and to sail at night. They made primitive charts and knew a crude form of dead reckoning. They used observations of the sun and the North Star, or polestar, to determine directions. They estimated distances from the time it took to cover them. Phoenicians dominated sea trade in the Mediterranean Sea for 3000 years.

Traditional navigators of the Central Caroline Islands in the Western Pacific developed skills to meet the needs of ocean voyaging for distances of up to several hundred miles among the tiny islands and atolls of Micronesia. Lacking writing, local navigators had to memorize their knowledge of the stars, sailing directions, seamarks, and how to read the waves and clouds to determine currents and predict weather. Before Europeans entered Micronesia, the known world of Carolinian navigators extended from Palau and Yap in the west to Ponape in the east and from Saipan and

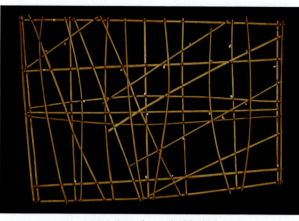

Navigation chart (rebbelib),
© 2008 The Trustees of the British Museum.

Guam in the north to Nuquoro and Kapingamarangi in the south. Their sailing directions also included places beyond this region in the west, south, and east, but these lay outside the limits of intentional voyaging and were mostly known as mythical rather than real places. Knowledge of such distant places met no practical need but served to show off one's learning.[15]

Ideas of trigonometry apparently were first developed by the Babylonians in their studies of the motions of heavenly bodies. About 70 Babylonian tablets have been found, originally from the second millennium BC, that refer to the appearances of the sun, moon, and planets, as well as meteorological phenomena. Because of the importance of celestial phenomena for the understanding of events in Babylonian society, the Babylonian temple astronomers had been observing the skies for centuries and had recorded their observations in so-called astronomical diaries, astronomical catalogs of stars and other texts from the seventh century until the first century BC. This is by far the longest

Islamic astrolabe (Whipple Museum, Cambridge University).

Ptolemy's system from a medieval Islamic manuscript.

continuous scientific record that has ever existed. Compare that with our modern science, which has existed for only half as long. The final, mathematical, phase of Babylonian astronomy dates mainly from the third to the first centuries BC. From this period, we have the ephemerides: tablets containing tables of the computed positions of the sun, moon, or planets, day by day, or over longer periods, such as month by month. There are also tablets called procedure texts, which give schematically the rules for computing ephemerides, much like a modern computer program. Our zodiac was also developed in Babylon.

In China, a calendar had been developed by the fourth century BC. A Chinese astronomer, Shih Shen, drew up what may be the earliest star catalog, listing about 800 stars. Chinese records mention comets, meteors, large sunspots, and novas, which mean they did extensive observations and data collecting.

Even Euclid wrote an astronomical work, *Phaenomena*, in which he studied properties of curves on a sphere, using *spherical* geometry, which is different from the geometry on a plane that we now call *Euclidean* geometry.

Model explaining Ptolemy's system
(Whipple Museum, Cambridge University).

Observations of heavenly bodies were carried out in ancient Egypt and Babylon, mainly for astrological purposes and for making a calendar, which was important for organizing society. Claudius Ptolemy (ca. 100–178 CE), in his *Almagest*, cites Babylonian observations of eclipses and stars dating back to the eighth century BC. The Babylonians originated the notion of dividing a circle into 360 degrees—speculations as to why they chose this number include that it was close to the number of days in a year, it was convenient to use in their sexagecimal system of counting, and 360 is the number of ways that seven points can be placed on a circle without regard to orientation (for the ancients, there were seven "wandering bodies"—sun, moon, Mercury, Venus, Mars, Saturn, and Jupiter).

The ancient Greeks became familiar with Babylonian astronomy around the fourth century BC and developed spherical geometry, as will be discussed in Chapter 5.

Navigation and large-scale surveying developed over the centuries around the world, and along with it cartography, trigonometry, and spherical geometry. Examples most closely associated with this strand in the last two centuries are the study of surfaces and manifolds, which led to many modern spatial theories in physics and cosmology.

Motion/Machines Strand

The first uses of some kind of mechanical device probably were log rollers placed beneath a load to be moved, as in the Paleolithic era (15,000–75,000 years ago). Based on a diagram found on ancient clay tablets, the earliest known use of the wheel was a potter's wheel that was used at Ur in Mesopotamia as early as 3500 BC. It is possible that there was an independent discovery of the wheel in China around 2800 BC, but there has been less historical evidence for this.

Etruscan pots produced on a potter's wheel.

The first use of the wheel in transportation was in Mesopotamian chariots around 3200 BC. A wheel with spokes first appeared in Egyptian chariots around 2000 BC, and wheels seem to have developed in Europe around 1400 BC without any influence from the Middle East. Celtic chariots introduced an iron rim around the spiked wheel, and this design, still unchanged, is used in horse carriages today! Despite the overwhelming utility of the wheel, some civilizations failed to discover it, for example, those of sub-Saharan Africa, Australia, and the Americas. Archeologists have found some children's toys from the Incan civilization that suggest that this society was at least familiar with wheel-like shapes, but they apparently were not used for utilitarian purposes.

Recently, scholars from Harvard University and the Max Planck Institute for the History of Science in Berlin analyzed technical treatises and literary sources dating back to the fifth century BC and found several mechanisms in use among practitioners with limited theoretical knowledge. For example, they found that the steelyard—a balance with unequal arms—was in use as early as the fourth and fifth centuries BC, before Archimedes and other thinkers of the Hellenistic era gave a mathematical explanation of its use, using the law of the lever.

In ancient Greece, Archimedes, Heron, and other geometers used linkages (straight sticks pinned together in a way that they can move) and gears (wheels with pins) to solve geometric problems, such as trisecting an angle, duplicating a cube, and squaring a circle (finding a square with the same area as a given circle). These solutions were not accepted in the building/structures strand, which leads to a common misconception that these problems are unsolvable and/or that Greeks did not allow motion in geometry. The truth is that one cannot solve these problems using only a compass and unmarked straightedge sequence.[16]

A linkage for trisecting angles.

Why did solvability of these three problems become so important in geometry? They really did not have such a big importance in practical applications. They are also not fundamental problems—there is no particular theory based on them. These three ancient problems became famous just because so many people tried to solve them and these attempts actually led to many new methods in mathematics.

It seems that motion was first explored in connection with astronomy (the geometry of the heavens), where planetary motion was translated into geometric terms so that techniques of Euclid's *Elements* could be applied. About 365 BC, the Greek scholar Eudoxus visited Egypt, where he acquired from the priests of Heliopolis knowledge of planetary motions and Chaldean astrology. Later he completed his book *On Speeds* about motions within our solar system (perhaps his greatest, but lost, writings). Eudoxus became the first mathematician to seriously attempt to describe the intricate motions of celestial bodies using a mathematical model based on spherical geometry. Geometry and motion came closer together for ancient engineers. According to some ancient references, one of the first mechanical solutions to the three famous problems was offered by Menaechmus (ca. 380 BC–320 BC), math tutor to Alexander the Great, but there are no actual accounts of it available. Around 420 BC, Hippias devised the mechanism that would draw a curve, called a *quadratrix*, which can be used to trisect angles and square circles.

Plato criticized this mechanistic approach and called instead for a purely theoretical solution. "Motion" would involve mechanics and experiments; implying manual work, but in ancient Greece it meant "fit only for slaves." As Aristotle (384 BC–322 BC) wrote: "These inferior persons should never be admitted to citizenship because no man can practice virtue that is living the life of a mechanic or laborer."

Still, the oldest known engineering textbook is attributed to Aristotle, though some authors think that it was written by his student Straton. In this book, we find the first mention of gear wheels. The Romans and Greeks made wise use of gearing in clocks and astronomical devices. Gears were also used to measure distance or speed. One of the most interesting relics from ancient Greece is the Antikythera mechanism, which is an astronomical computer. It had many gears in it, some of which were planetary gears. Just before the opening of the 2008 Olympic Games, scientists announced the discovery that this mechanism was a complex clock that combined calendars and also showed the four-year cycle of the Ancient Greek games.[17]

The greatest geometer and engineer of antiquity was Archimedes (ca. 282 BC–ca. 212 BC). Plutarch wrote about him:

> He would not deign to leave behind him any writings on his mechanical discoveries. He regarded the business of engineering, and indeed of every art which ministers to the material needs of life, as an ignoble and sordid activity, and he concentrated his ambition exclusively upon those speculations whose beauty and subtlety are untainted by the claims of necessity…. Certainly in the whole science of geometry it is impossible to find more difficult and intricate problems handled in simpler and purer terms than in his works.[18]

In fact, however, there is evidence that Archimedes did write on certain mechanical subjects, for example, his book *On Sphere Making* or *The Method*—a discovery by mechanics of many important results about areas and volumes.

The Antikythera mechanism.

The trammel as a children's toy.

Contemporary auger at a construction site that is the same Archimedean screw still used after more than 2000 years.

Planetary gear or Tusi couple mechanism, Kinematic Model Collection, Cornell University (photo by Prof. Francis Moon).

Since ancient times, mechanisms were used for drawing curves. For example, the trammel is the simplest mechanism for drawing ellipses. It was described by Proclus, but it is also attributed to Archimedes.

Al-Tusi (1201–1274) was among the first of several Muslim astronomers who found several serious shortcomings in Ptolemy's planetary model based on mechanical principles and modified it. He devised several instruments for astronomical observations, but his best known device is the so-called Tusi couple. Later, this mechanism was called in kinematics "a planetary motion mechanism" and was used as one of the straight-line mechanisms in order to convert circular motion into straight-line motion in machines (see discussion below).

The philosophical approach to the description of motion mathematically was done in the thirteenth century in the so-called *Merton School* by Thomas Bradwardine (ca. 1290–1349) and others. In the fourteenth century, motion was discussed in writings by Jean Buridan (ca. 1300–ca. 1358) and Nicole Oresme (ca. 1320–1382). Oresme represented motions geometrically by plotting primitive graphs. Motion and mathematics were important objects of interest in research done by Tycho Brahe (1546–1601) and Galileo Galilei (1564–1642).

One of the most significant turning points in the development of technology was learning how to transform continuous circular motion into reciprocal or straight-line motion. Rotary motion was available to humans using natural forces—waterwheels, windmills. But this kind of motion was not enough—for example, to saw logs into boards, rectilinear motion was needed. This transformation was achieved by the use of gears and linkages. Both later became important subjects of mathematical interest.

Watermill in Braine-le-Château, Belgium,
dating from the twelfth century.

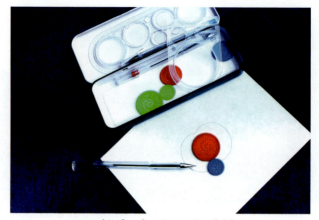

Toy kit for drawing epicycloids.

To construct the most efficient shape of gear teeth, geometers were studying cycloids (Nicolas of Cusa in 1451, Galileo in 1599) and epicycloids (Albrect Dürer in 1525). Apollonius and Ptolemy had discussed the motion of planets in geometric terms, and that is where a mention of epicycloids appears for the first time.

In 1557, Girolamo Cardano first published a mathematical theory of gears. In 1694, Philippe de la Hire published a full mathematical analysis of epicycloids and recommended involute curves for gearing, but in practice this was not used for another 150 years. In 1733, Charles Camus expanded la Hire's work and developed theories of mechanisms. In 1754, Leonhard Euler worked out design principles for involute gearing.

Another mechanism based on geometric proportions and known since ancient times is a *pantograph*. It can be called the earliest copying machine, making exact duplicates of written documents. Artists adopted pantographs for duplicating drawings and enlarging sketches. One of them was Leonardo da Vinci, who used a pantograph to duplicate his sketches on canvas. Later pantographs were adopted specifically for duplicating paintings—first the pantograph would be used to trace the outlines and then the shapes would be filled in with paint. Sculptors and carvers adopted pantographs for tracing master drawings onto blocks of marble or wood. In the eighteenth century, the pantograph was used to cut out typeset letters for printings and engravings. In the nineteenth century, pantographs were advanced enough to duplicate sculptures. One of the first such duplicated sculptures was Michelangelo's sculpture *David*. Heavy-duty pantographs are still used for engraving and contour milling.

Leonardo da Vinci had ideas about several mechanisms that would trace various mathematical curves. Mechanical devices for drawing curves were also used by Albrecht Dürer.

Pantograph (Diderot and d'Alambert Encyclopedia, 1755-1780).

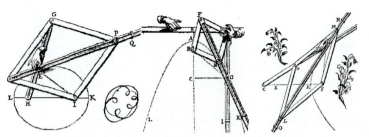

Schooten's illustrations for Descartes' *Geometry*.

By the beginning of the seventeenth century, mathematicians had developed a new "language" for representing various arithmetic concepts and relationships: symbolic algebra. Geometry, however, was still considered as the more trusted form of expressing mathematical thought, partially due to the tradition of authority of Euclid's *Elements*, where even the concepts of number theory were expressed in geometric form. The scientific revolution prompted experiments in representing geometric concepts and constructions in this new symbolic language. In seventeenth-century Europe, questions of appropriate forms of representation were dominant intellectual activities. They appeared not only in mathematics and physics but maybe even more in religious, political, legal, and philosophical discussions.

Therefore, it is not surprising that Descartes and Leibniz were paying so much attention to symbolic representations of their mathematical ideas; they viewed these investigations as part of their extensive philosophical works. Descartes' *Geometry* was originally published as an appendix to his philosophical work *Discourse of the Method*. Political thinkers of the time, like Thomas Hobbes, commented on the latest developments in mathematics and physics. Descartes' *Geometry* is considered to be the start of analytic geometry—using algebraic methods for solving geometry problems. But nowhere in his book had he written symbolic equations. He studied curves that were constructed by mechanical devices. After the curves had been drawn, Descartes would introduce coordinates, analyze the motion that resulted in the particular curve, and arrive at an "equation" written out as a sentence that would represent this curve. Curves were creating equations and not the other way around—the way we are used to studying curves today. Descartes used equations to create a taxonomy of curves. He could assume his audience would be familiar with Euclid's *Elements* and Apollonius' *Conic Sections*.

Descartes was not alone: independently, Fermat came up with ideas of analytic geometry. Roberval, Wallis, Cavalieri, and Newton all tried to express their geometric findings about motion in symbolic language. These efforts culminated in creating calculus by the late seventeenth century.

There was an interaction between mathematics and mechanics that led to marvelous machine design and that continues to the modern mathematics of rigidity and robotics. For more information and history about the interactions between mathematics and mechanisms, see the website KMODDL.org.

Chapter 4

Tidbits from the
History of Crochet

Since in this book crochet is used as a medium to create mathematical models, let us have a look at some history of crocheting, too.

Artifacts from ancient times that can tell a story about the use and techniques of fibers are less preserved than, say, pieces of pottery, coins, or jewelry. Still, scholars have been able to find fragments and put, at least partially, the story together. Lila de Chaves, a textile scholar from Greece, has concluded that interlacing with loops was used by the people of the Neolithic Age before the discovery of the spindle wheel, or spinning wheel, as far back as 5000 BC. Those pieces are referred to as "knotless netting" or "naalebinding."[1]

Naalebinding is a technique for making fabric with a needle and yarn with looping. One of the most ancient samples of naalebinding came from Egypt between the fourth and sixth centuries BC. It is generally agreed that

Naalebinding or "shepherd's knitting" (National History Museum of Latvia).

Naalebinding.

naalebinding was a prelude to needlepoint lace and interlacing and a forerunner to knitting and crochet.[2]

There is no particular time to which the beginning of the handiwork technique called crochet is dated. Mary Konior in her book[3] quotes Heinz Edgar Kiewe (1906–1986) as stating that the earliest written reference to crochet hooks is dated between the years 50 and 137 CE, but there are no sources given for this. The earliest knitted fragments discovered in archeological excavations are from the second to seventh centuries.

Kiewe was born in Königsberg, East Prussia. He was a self-styled "textile journalist" who ran a needlework/yarn shop called Art Needlework Industries Ltd. in Oxford, England, from about 1940 until the late 1980s. One of the books he self-published was *The Sacred History of Knitting*.[4] In this book, he also remarks on the sacred history of crochet and suggests that it was practiced in the Near East before the days of Solomon.[5] Konior thinks that that may well be an overestimate but notes that it is generally accepted that both knitting and crochet evolved in the early Arabic civilizations and then spread through Arab traders. Kiewe also concludes that the history of the ancient arts of crochet, knitting, netting, weaving, twisting, and braiding are so closely related that they may be inseparable. Other suggestions are that crochet probably developed most directly from Chinese needlework, a very ancient form of embroidery known also in Turkey, India, Persia, and North Africa, which reached Western Europe in the eighteenth century and was called *tambouring*—from the French *tambour* (drum). Tambouring was a technique in which thin background fabric was stretched taut on a frame, with the thread kept underneath this fabric, and then a needle with a hook was inserted downward and a loop of the working thread drawn up through the fabric. Later this underneath fabric was discarded.

In Latvian, which is one of the oldest living Indo-European languages, the name for crochet comes from this tambouring stem. The oldest mitten found in Latvia has been dated from the thirteenth century; it was created in "shepherd's knitting."

The first knotted mittens with ornaments and designs are dated to the fifteenth century though knitting needles that were found already in fourteenth-century archeological diggings. The National History Museum of Latvia has a collection of about 3000 mittens; this collection was started in the nineteenth century.

Latvian mittens and socks with traditional ornaments.

It is interesting to find that in different places, handiwork developed independently. For example, as described by K. Bühler,[6] the Warrau Indians of Venezuela have their techniques of crochet and knitting that differ widely from any other known methods and therefore could not have been acquired from an outside source.

There has always been a strong tradition of crochet among Muslim populations of the Middle East, and it is also evident in tribal cultures of central and northern Africa—in fact, there is a crochet technique named Tunisian crochet, which employs one long crochet hook.

An old crochet pattern book
from the collection of Gwen Blakley Kinsler
(photo courtesy of Interweave Press).

crochet technique, in Western Europe crochet was used to partially replace more expensive and laborious lace making. Paludan offers three theories based on the early crochet evidence:

- it is possible that crochet originated in Arabia and spread eastward to Tibet and westward to Spain;

- earliest evidence of crochet might be seen in some adornments in rites of puberty for primitive tribes from South America;

- three-dimensional dolls have been found in Ancient China made with crochet technique.

A fairy tale by H. C. Andersen called "The Girl Who Trod on the Loaf of Bread" (first printed in 1859) describes an old crone of the marshes who embroidered lies and crocheted thoughtless words. Maybe that was an echo of some attacks in real life on such apparently harmless pursuits as crochet. In the second half of the 1800s and the beginning of the 1900s, crochet was banned from needlecraft classes in Prussian schools on the grounds that it was a superfluous pastime, unlike learning useful skills such as how to patch, darn, knit, and sew linen. Teachers protested, and one of them, R. Schallenfeld, in her 1861 book argued that crochet was the very needlecraft best suited for teaching children all the rules of working with yarns (including knitting), as well as being the best way to cultivate their sense of form. She pointed out that of all yarn-working techniques, crochet provided the best opportunity for self-development and creativity.[8]

Annie Potter calls this lack of evidence and disagreement about early crochet roots "a living mystery."[9] In her book, she claims that the modern art of crochet emerged as a result of experimentation with needle and thread during the Renaissance period in Italy and France and quickly

Crochet is an easy work to lay aside and take up again, for there is only one hook and one stitch at a time, and thus it is easy to set down in an emergency. Perhaps this convenience for travel accounts for the popularity of crochet among canal-boat and gypsy communities.

Lis Paludan did extensive research on the history of crochet, and she found that it was impossible to find evidence of crochet, as it is known now, before 1800 in Western Europe.[7] She wrote that early forms of crochet suggest that evidence of it should be sought in two different environments: partly in the poor, cold regions of northern and southeastern Europe, and partly in the prosperous milieu of Western Europe. This evidence can also be seen in crochet techniques: while in cold places crocheted pieces were mostly durable mittens made with a simple fabric-like spread to other countries such as Flanders (Belgium), England, Ireland, and America. As one of the earliest written documents mentioning crochet she points to the "Letters Patent" granted to the Mercers of France in the year 1653 that described the enrichment of all forms of braid and lace, including "au crochet, et au fuseau."[10] Potter also writes:

> Anyone familiar with the skills of crocheting has experienced the freedom that the art form offers. Its variety and flexibility challenge the imagination and there is quite literally no known boundary for what can be accomplished—with a simple hooked tool and a continuous strand of material.[11]

The therapeutic properties of crochet have also been recounted, like in this suggestion from ca. 1916:

> Of recent times medical authorities have declared that crochet work, undertaken in moderation, is one of the most beneficial of home hobbies, as it not only occupies the fingers, but the mind, and is soothing to the nerves. One eminent physician has gone so far as to recommend the pastime to men as well as to women, and to suggest that it might be regarded as a sedative and a substitute for a smoke.[12]

Whatever is the history, it is clear that crochet technique is quite old and widespread, as is knitting.

In the late 1960s, a few artists in the United States began to use crochet as a vehicle for artistic fabric or as an art form. Clinton D. MacKenzie wrote:

> With crochet, one can be as systematic as an architect assembling planes and as free as a poet describing his train of thought. Size and design possibilities are limited only by one's imagination and energies.... Crochet gives form, structure, mass, scale, and meaning to the thread. It is a simple process, but it is flexible enough to produce a work of art when the construction follows basic aesthetic concepts. With crochet you can write your own story line, and make your fantasies grow from toys to monuments.[13]

Recently, free-form crochet has become very popular all over the world. Crafters are using a hook as a brush and their yarn as paint, but instead of two-dimensional paintings, with crochet, artists can make their creations three-dimensional. Free-form crochet means that one is not following somebody else's pattern but creates an original work.

Knitting as a medium to explain mathematical ideas was used already in the nineteenth century. In the Whipple Science Museum at the University of Cambridge, there are two surfaces knitted by Professor Alexander Crum Brown (1838–1922). Already as a child he was very interested in models. Before he reached school age, he created a small machine for weaving a cloth, thus starting his lifelong hobby with knots and knitting. The models on display in

Mathematical surfaces knitted by Alexander Brown (Whipple Science Museum, University of Cambridge).

the Whipple museum were used in his lectures on crystallography when he was a professor in the University of Edinburgh.

An example of a crocheted mathematical surface is the Riemann surface in the mathematical model collection at the University of Göttingen. Unfortunately, I do not know anything about the history of this model nor the person who created it. The photo of it was published in *The Mathematical Intelligencer* in 1993.[14] I wish I had known this fact before publishing a description about my hyperbolic plane models in *The Mathematical Intelligencer*[15]—I would have had an argument that crocheted mathematical surfaces were featured in this publication already at least once!

Crocheters have been interested in turning flat things into three-dimensional things. An interesting example is ruffled doilies. B. J. Licko-Keel sent me a booklet about ruffled doilies.[16] She was curious whether these are already patterns for crocheted hyperbolic planes. Increases in these patterns do not keep a constant ratio, so these surfaces do not have *constant* negative curvature, but they certainly have negative curvature.

Crocheted Riemann surface
in the mathematical model collection at the University of Göttingen
(photo by Elisabeth Mühlhausen).

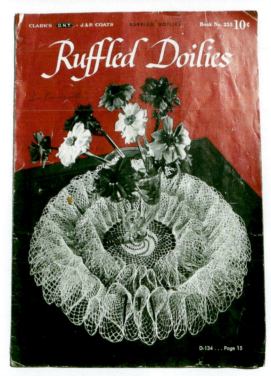

Example of crocheted surface with negative curvature
(courtesy of B. J. Licko-Keel).

Chapter 5

What Is Non-Euclidean Geometry?

There seems to be much confusion about the history of non-Euclidean geometries. Let us try now to look through the lenses of strands in the history of geometry (Chapter 3) to see how non-Euclidean geometries developed. We have talked already a little bit about how mathematicians started to think about the hyperbolic plane and a new type of geometry in the nineteenth century. Often this is considered to be the beginning of non-Euclidean geometries. But is it really true that non-Euclidean geometry was first discovered only in the nineteenth century? Euclid himself studied geometry on different surfaces!

Non-Euclidean Geometries in the Navigation/Stargazing Strand

Babylonians had recorded astronomical observations since the eighth century BC. The ancient Greeks became familiar with Babylonian astronomy around the fourth century BC. One of the greatest ancient Greek scholars, Eudoxus (408–355 BC), developed the *two-sphere model* for astronomy. In this model, he considered the stars to be on the celestial sphere (which rotates one revolution a day westward about its pole, the North Star) and the sun to be on the sphere of the ecliptic. On this second sphere, he considered the equator as the path of the sun; the angle between it and the equator of the celestial sphere was about 24 degrees in Eudoxus' time. (Today this angle has changed and is about 23.5 degrees.) The sphere of the ecliptic has an apparent rotation eastward of one revolution a year. Both of these spheres appear to rotate about their own poles.

Another ancient Greek scholar, Autolycus (ca. 360 BC–ca. 290 BC), in his manuscript *On the Rotating Spheres* (written sometime between 333 and 300 BC), introduced a third sphere whose pole is the point directly overhead a particular observer and whose equator is the visible horizon. This book and Euclid's *Phenomena* (300 BC) are the earliest known mathematical works that mention spherical geometry. Both of these books use theorems from spherical geometry to solve particular astrological problems, for example, to answer the question of what the length of daylight is on a particular date at a particular latitude.

The first systematic account of spherical geometry was *Sphaerica* by Theodosius (around 200 BC). It consisted of three books of theorems and construction problems.

Armillary sphere based on Autolycus' three-sphere model (Vatican museums).

Within the navigation/stargazing strand, we can realize that spherical geometry was the first non-Euclidean geometry. Euclid himself was considering geometry on surfaces with intrinsic geometry different from the traditional plane geometry (what we call Euclidean geometry). In Chapter 2, we gave some examples of how geometry on a sphere differs from geometry on a plane.

Carl Friedrich Gauss (see Chapter 1) in the nineteenth century developed the notions of curvature and constant curvature in the context of his surveying the surface of the Earth. He used theorems of spherical geometry in this surveying work.

Armillary sphere (Science Museum, Milan).

Non-Euclidean Geometries in the Building/Structures Strand

What we now call Euclidean geometry was developed in Euclid's *Elements*, compiled around 300 BC. It was a collection of Greek knowledge of geometry in plane and three-dimensional space, which we considered within the building/structures strand in Chapter 3. It should be mentioned that *Elements* was a geometric treatment also of other mathematical ideas besides geometry itself. This work consisted of 13 books, and some of them were about algebra and number theory. Euclid's work became a model of how mathematical reasoning can be carried out. This presentation was so successful that for the next 2000 years, the *Elements* was the most studied book in mathematics. These studies affected some people very personally. For example, one of the most important logicians of the twentieth century, Bertrand Russell (1872–1970), remembered:

> At the age of eleven, I began Euclid, with my brother as my tutor. This was one of the great events of my life, as dazzling as first love.[1]

Euclid did not pretend that he was describing physical space. In those days, physical space was Aristotle's universe and was believed to be a finite one. Aristotle divided the universe into a sublunary sphere of daily life, the region of the planets, and a sphere of fixed stars.

Still, there was a puzzling thing in Euclid's *Elements* that drew mathematicians' interest for many centuries. It was the so-called Parallel Postulate (also called the Fifth Postulate):

> If a straight line intersecting two straight lines makes the interior angles on the same side less than two right angles, then the two lines (if extended indefi-

nitely) will meet on that side on which are the angles less than two right angles.

Neoplatonist philosopher and mathematician Proclus (412–485 CE) was one of the first who raised eyebrows about the Fifth Postulate. He wrote:

> [Euclid's Fifth Postulate] ought to be struck from the postulates altogether. For it is a theorem—one that invites many questions … and requires for its demonstration a number of definitions as well as theorems … it lacks the special character of a postulate.[2]

During and since the Greek era, there were many attempts to derive the Parallel Postulate from the rest of elementary geometry, attempts to reformulate the postulate or the definition of parallels into something less objectionable, and descriptions of what geometry would be like if the postulate was in some way denied. Part of these attempts was to study which geometric properties could be proved without using the Parallel Postulate but with the other postulates (axioms) of the plane. The results of these studies became known as *absolute geometry*.

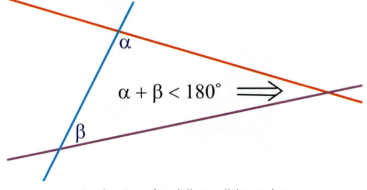

Visualization of Euclid's Parallel Postulate.

During the ninth through twelfth centuries, the Parallel Postulate in different forms was explored by mathematicians in the Islamic world. Great medieval scientist Abu Ali al-Hasan ibn al-Haytham (965–1039), who is perhaps better known in the Western world as Alhazen, was born in Basra, Iraq. He proved the Parallel Postulate by assuming that *a quadrilateral with three right angles must have all right angles*. Quadrilaterals with three right angles were known later in Western literature as Lambert quadrilaterals after Swiss German mathematician, physicist, and astronomer Johann Lambert (1728–1777), who studied them. Lambert also was the first to extensively investigate hyperbolic trigonometric functions.[3] The first proof that π is irrational was done by Lambert.

The Persian poet and geometer Omar Khayyám (1048–1122) wrote a book that in translation is entitled *Discussion of Difficulties in Euclid*, in which he introduced a new postulate, which says that *two straight lines that start to converge continue to converge*. In his work on parallel lines, he studied the Khayyám quadrilaterals, which have two right angles adjacent to the same side (later in the West called Saccheri quadrilaterals).

Nasir al-Din al-Tusi (1201–1274) furthered the study of parallel postulates and is credited (though some say it might have been his son) with first proving that Euclid's Parallel Postulate is equivalent to the assumption that the *sum of interior angles in a triangle in a plane is 180 degrees*. Al-Tusi's works were the first Islamic mathematical works to be discovered in the Western Renaissance and were published in Rome in 1594. The assumption that *parallel lines in a plane are equidistant* was discussed in various forms by Aristotle, Posidonius (135–51 BC), Proclus, ibn Sina (980–1037), Omar Khayyám, and Saccheri (1667–1733). For more discussions, see Chapter 10 of *Experiencing Geometry*.[4]

One of the contributors to the development of calculus, English mathematician John Wallis (1616–1703), proved that Euclid's Parallel Postulate follows from the assumption that *to every triangle, there exists a similar triangle of arbitrary magnitude*. Those were not the only attempts to rephrase the Parallel Postulate or substitute it with something else. A German doctoral thesis by Georg Simon Klügel written in 1763 recorded 28 attempts to prove the parallel postulate.[5] After the seventeenth century, these investigations in geometry were continued in the French school. Early nineteenth-century French mathematician and physicist Joseph Fourier (1768–1830) concluded that geometry was a physical science and could not be established a priori. Adrien-Marie Legendre (1752–1833) proved that in absolute geometry, the sum of the interior angles of a triangle is always less than or equal to 180 degrees.

Some mathematicians became irritated with all these attempts to prove the Parallel Postulate. For example, Pierre-Simon Laplace (1749–1827), who was the leading applied mathematician of his time, said that Newton's law of gravitation is so well attested by experience that it must be regarded as true. Newton's law permits arbitrary scalings, and so the existence of similar, noncongruent figures must be allowed, and thus (by Wallis' result) the parallel postulate is true. This is, at best, an argument that the world we live in can be described approximately by Euclidean geometry; it does not show that the Parallel Postulate is necessarily true.[6]

It is difficult to know what Carl Friedrich Gauss (1777–1855) was thinking about the Parallel Postulate, because he never put all his ideas about it together. (His motto was not to publish something unless it was complete.) We can only rely on scattered remarks in his correspondence. In 1817, for example, he wrote to the astronomer Wilhelm Olbers:

> I am becoming more and more convinced that the necessity of our geometry cannot be proved … . Perhaps only in another life will we attain another insight into the nature of space, which is unattainable to us now. Until then we must not place geometry with arithmetic, which is purely a priori, but rather in the same rank as mechanics.[7]

In one of his letters to F. A. Taurinus, Gauss had set out his views on possible non-Euclidean geometry but asked Taurinus to keep them confidential:

> The assumption that the sum of the three angles [of a triangle] is smaller than π leads to a geometry which is quite different from our [Euclidean] geometry, but which is in itself completely consistent. I have satisfactorily constructed this geometry for myself so that I can solve every problem, except for determination

Carl Friedrich Gauss
(painting by Gottlieb Biermann, 1887,
Göttingen University collection).

of one constant, which cannot be ascertained a priori. The larger one chooses this constant, the closer one approximates Euclidean geometry.[8]

We can only speculate that perhaps Gauss was concerned about his fame as "prince of mathematics" and that he did not want his name to appear in discussions about doubtful ideas.

The breakthrough in the study of parallel postulates came in the nineteenth century when, apparently independently, János Bolyai (1802–1860) and N. I. Lobachevsky (1792–1856) developed a new geometry that was an absolute geometry that did not satisfy Euclid's Parallel Postulate. It was to this new geometry that the label "hyperbolic geometry" was later attached.

János Bolyai's father, Farkas Bolyai (1775–1856), studied mathematics at Göttingen University, and one of his fellow students was Carl Gauss. When Farkas returned back to his native Hungary in 1798, he kept in touch with Gauss and continued to discuss the Parallel Postulate. By the age of only 12, his son, János Bolyai, read the first six books of Euclid's *Elements* and started to attend lectures in college. János graduated from college in 1817 with prizes in Latin and was also an acclaimed violinist. From 1818 to 1823, János Bolyai studied at the Royal Engineering Academy in Vienna, which trained cadets for military service. He then served as an engineer in the Austrian Army for ten years, and in 1833 he retired on a pension as a semi-invalid.[9] By 1820, he started to think about the Parallel Postulate. His father was worried that his son was going in a dead-end direction and wrote to him:

> You must not attempt this approach to parallels. I know this way to the very end. I have traversed this bottomless night, which extinguished all light and joy of my life. I entreat you, leave the science of parallels alone … I thought I would sacrifice myself for the sake of the truth. I was ready to become a martyr who would remove the flaw from geometry and return it purified to mankind. I accomplished monstrous, enormous labours; my creations are far better than those of others and yet I have not achieved complete satisfaction… I turned back when I saw that no man can reach the bottom of this night. I turned back unconsoled, pitying myself and all mankind. Learn from my example: I wanted to know about parallels, I remain ignorant, this has taken all flowers of my life and all my time from me.[10]

The son did not listen to his father and in 1823 he wrote to him:

> All I can say now is that I have created a new and different world out of nothing.

Statue of János and Farkas Bolyai, Marosvásárhely, Romania (sculpture by C. I. Márton, 1957).

The father advised the son to publish his results as soon as possible and offered to include them as an appendix to the geometry book he was working on. It was published in 1832. A copy of this book was sent to Gauss, who wrote to Farkas Bolyai:

> To praise it, would be to praise myself. Indeed the whole contents of the work, the path taken by your son, the results to which he is led, coincide almost entirely with my meditations, which have occupied my mind partly for the last thirty-five years. … up till now I have put little on paper, my intention was not to let it be published during my lifetime.[11]

Farkas was pleased to hear back from the great mathematician who indirectly was praising his son's work, but János Bolyai felt insulted. There was a long period when father and son did not speak to each other: the son because he suspected that his father had in some previous correspondence conveyed his ideas to the old friend, and the father because he disapproved of the fact that János lived unmarried with a woman by whom he had three children.

The immediate reception of Bolyai's work was poor—nobody really noticed an appendix to a two-volume work written in Latin, which was still the tradition for writing scientific works. Gauss had a chance to draw the attention of the mathematical world to this discovery, but he failed to promote Bolyai's discovery for reasons known only to him. The result was that Bolyai's discovery passed unnoticed by the international mathematical community in the lifetimes of both the father and the son.

János Bolyai wanted to be the first to discover the new geometry because, as he wrote:

> …the ideas might easily pass to someone else who would then publish them. … there is some truth in this, that certain things ripen at the same time and then appear in different places in the manner of violets coming to light in early spring. And since all scientific striving is only a great war and one does not know when it will be replaced by peace one must win, if possible; for here pre-eminence comes to him who is first.[12]

Every mathematician would be happy to open the doors to a new world in geometry. We already saw that Gauss was thinking a long time about the possibility of a new geometry. But János Bolyai was right: similar ideas do appear independently at the same time in different places once the idea is ripe. The case of discovering this non-Euclidean geometry was one of them. While János Bolyai celebrated his discovery, the same joy was celebrated by the

Nikolai Ivanovich Lobachevsky
(painting by Lev Krjukov, University of Kazan).

Russian mathematician Nikolai Ivanovich Lobachevsky (1792–1856). But regrettably his work had a similar fate as Bolyai's—it also went almost unnoticed.

Lobachevsky first reported his results on a new geometry in 1826. In 1829, he published the first of several articles in the *Kazan Messenger* describing an alternative geometry to Euclid's. Lobachevsky showed that in his new geometry, the angle sum of the triangle is always less than two right angles, and the angle sum gets less as the triangle gets bigger. Lobachevsky never showed that his new geometry was consistent. In 1837, Lobachevsky published his article "*Géométrie imaginaire,*" and a summary of his new geometry, *Geometrische Untersuchungen zur Theorie der Parellellinien*, was published in Berlin in 1840.

Connecting Constant Curvature with Non-Euclidean Geometry

The notion of curvature and constant curvature grew out of the navigation/stargazing strand. Euclid's *Elements* and the study of its Parallel Postulate grew out of the building/structures strand. Because of this, it is not surprising that the connections were difficult to see. Examples of spaces with negative curvature had been given in the literature, but no one had connected them to non-Euclidean geometry.

After Gauss died in 1855, mathematicians started to work through his many notebooks and other papers. In the case of hyperbolic geometry, they found many notes about it among his papers, as well as the books by Bolyai and Lobachevsky. J. Gray writes:

> Also his letters confirmed that for many years Gauss had thought that a new geometry was possible. He

refrained from publishing either because the right idea never came to him (not even through the post) or because he was truly afraid of the clamor such a discovery would set off, or perhaps that priority had already gone to someone else. It was indeed the case that neither he, nor the Bolyais, nor Lobachevsky, had precisely shown that a contradiction would never be found.[13]

Apparently, Gauss also thought about the possibility of having a surface of constant negative curvature, but he never published his thoughts. The mathematician who first had the idea that Gauss "missed" was a student of his, Bernhard Riemann (1826–1866).

In 1854, Riemann was presenting his *Habilitationschrift* at Göttingen University to get the qualification permitting him to teach. Rules required the candidate to offer three possible topics for his lecture. At that time, Riemann had already defended his PhD thesis (1851), in which he studied the theory of complex variables and introduced topological methods in complex function theory. Gauss described Riemann's thesis as "gloriously fertile originality."

For his inaugural lecture, Riemann submitted three topics. He had done groundbreaking work about the first topic and expected this to be the one chosen; the second topic was in another area where Riemann was an expert; and the third one was on his interest in geometry, but he had no previous publications on it.

Gauss was to choose which of the lectures should be presented, and to the surprise of Riemann, he chose the third one: *Über die Hypothesen welche der Geometrie zu Grunde liegen* (On the hypotheses that lie at the foundations of geometry). Riemann had about eight weeks to prepare for the lecture. He wrote to his father that he was convinced that Gauss had worked on this subject for years himself.

Bernhard Riemann
(University of Hamburg collection).

As a starting point, Riemann took one of Gauss' best results, *Theorema Egregium* (or Exceptional Theorem), which was one of the key results in Gauss' book on differential geometry, published in 1827. Gauss' result was that on a surface, there is a way of defining its curvature at each point that is intrinsic. Recall that this means that the curvature of a surface can be determined by taking measurements entirely on that surface, not stepping off of it into space. We talked about curvature in Chapters 1 and 2 and will discuss it more in Chapter 6.

Riemann's idea was to start with the concept of curvature and to argue that geometry was fundamentally about two types of problems: the intrinsic properties of a surface, and the ways in which a surface is mapped into another surface. Riemann did even more—he showed that this idea of Gaussian curvature can be generalized into higher dimensions. He also mentioned that there were three two-dimensional geometries that have constant curvature: it could be positive, negative, or zero (as we discussed in Chapter 1). In two-dimensional geometry, he proved that constant positive curvature implied locally spherical geometry, constant zero curvature implied locally Euclidean geometry, and constant negative curvature implied locally hyperbolic geometry. Marcel Berger has written more about Riemannian geometry.[14]

Non-Euclidean Geometries in the Art/Patterns Strand

Artists' experimentations with perspective in art led to the mathematical theory of *projective geometry*, which developed initially independently of both spherical and hyperbolic geometries. Then, in the latter part of the nineteenth century, projective geometry was used to provide a unified treatment of the three geometries. It was within this development that Felix Klein introduced the names *double elliptic geometry* (for the geometry of the sphere), *elliptic geometry* (for projective geometry), *parabolic geometry* (for Euclidean geometry), and *hyperbolic geometry* (for the Bolyai-Lobachevsky geometry).

Felix Klein (1849–1925) was a brilliant young mathematician when he was appointed full professor at the University of Erlangen at the age of 23, which was very unusual at that time. Like Bernhard Riemann, Klein worked in function theory, but he had a very strong interest in geometry. As was the tradition in German universities at the time, Klein also had to give an inaugural lecture, which he prepared but never actually gave. He published it,[15] and it became known as his Erlangen program. In this 1872 paper,

Felix Klein (University of Hamburg collection).

Felix Klein laid out a groundbreaking vision of geometry with a different emphasis than Riemann had in his inaugural lecture. Klein classified different geometries according to their symmetries (or groups of transformations); in this way, he showed a link between geometries and algebra.[16] Symmetries (and thus their associated algebras) came into geometry through the art/patterns strand.[17]

Further Investigations of Hyperbolic Geometry

Riemann's ideas circulated slowly in mathematical circles. He never saw them flourish because they were published only in 1868, two years after his death. The Italian mathematician Eugenio Beltrami was one of the first to follow them and to publish his own account on non-Euclidean geometry. He was also the first to find a physical surface on which this geometry, at least partially, would work.

Bolyai and Lobachevsky described the new geometry theoretically, but mathematicians searched for a surface on which the complete hyperbolic geometry would be true. In the next chapter, we will discuss how Beltrami came up with the idea of a surface with constant negative curvature, the *pseudosphere*, and we will show how to crochet it.

This surface represents only a little piece of the hyperbolic surface, so mathematicians continued to search for the surface that would be the complete hyperbolic plane. We talked about this in the Introduction and Chapter 1.

By the end of the nineteenth century, non-Euclidean geometry had become a widely discussed subject not only in mathematical circles. Since Euclidean geometry had been taught for many hundreds of years to every educated person, there were a large number of people who truly believed that Euclidean geometry is something like an absolute truth. For example, philosopher Gottlob Frege wrote:

> No one can serve two masters. One cannot serve truth and untruth. If Euclidean geometry is true, non-Euclidean geometry is false, and if non-Euclidean geometry is true, Euclidean geometry is false.[18]

Henri Poincaré originally discussed hyperbolic geometry in his 1895 essay "*L'espace et la géométrie*" and in some other writings. In 1902, these writings and other essays were compiled and printed in Poincaré's famous *La Science et l'hypothése*. This book became a real bestseller in France and was widely discussed. Linda Henderson[19] describes how these ideas influenced modern painters, for example, Francis Picabia and Marcel Duchamp.

Chapter 6

How to Crochet a Pseudosphere and a Symmetric Hyperbolic Plane

Pseudosphere

Eugenio Beltrami (1835–1900) came from an artistic family—his father was a painter who painted miniatures, and his grandfather engraved precious stones. Eugenio found that for him, music was more important, but in his career path he pursued his mathematical talent. Influenced by the works of Gauss, Lobachevsky, and Riemann, Beltrami considered the problem of geodesics on different surfaces— that is, what will be straight on different surfaces? He discovered that the surfaces with constant curvature will have geodesics that will be analogous to the behavior of straight lines in the plane. In 1868, he published a paper in which he showed the first concrete realization of Lobachevsky geometry: he showed that this geometry holds locally on the pseudosphere, the surface generated by the revolution of a tractrix about its asymptote.

The *tractrix* (from the Latin verb *trahere* meaning "to pull or drag") was studied by Dutch mathematician Christiaan Huygens (1629–1695), the first patent holder of a pendulum clock design, in 1692, who gave this name to the curve. The tractrix was of great interest to other famous mathematicians, such as the great German philosopher and mathematician, one of the inventors of calculus, Gottfried Wilhelm Leibniz (1646–1716); his devoted student, Swiss mathematician and physicist Johann Bernoulli (1667–1748), who belonged to one of the largest dynasties in mathematics; and others. This study was initiated by the following problem, posed as a challenge by Leibniz:

> What is a path of an object dragged along a horizontal plane by a string of a constant length when the end of the string not joined to the object moves along a straight line in the plane?

Leibniz called this the problem of the trajectory of a pocket watch, having heard it from Claude Perrault (1613–1688), who was a well-known physician, mechanic, and musician in Paris. Imagine that you have a pocket watch on a cord and you put it along an edge of a table as shown in the picture. Now, if you drag the end of the cord along the other edge of the table, the watch will move on a tractrix. The edge of the table that the cord moves along is the *asymptote* to the tractrix.

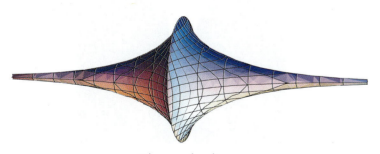

The pseudosphere.

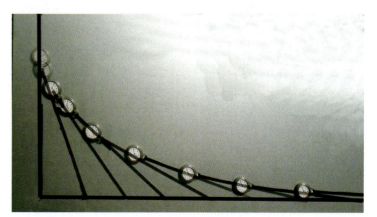

A pocket watch generating a tractrix.

Sometimes in calculus books this same curve is called "the lazy dog curve."

The pseudosphere can be used as the shape of a loudspeaker horn. In the early twentieth century, when gramophones were still a new technology, several designers carried out theoretical research in order to find the best shape for a horn. In 1927, British designer P. Voigt patented a horn design based on a tractrix. He based his assumption on the reasoning that if the curvature increases from plane waves (zero curvature) at the throat to a certain curvature at the mouth, then a point on an axis must travel at a faster rate than a point at the wall. Since the entire wavefront must travel at the speed of sound (assumed to be constant throughout the horn), the wavefront must be spherical, which means that it should have constant positive curvature. So, to complement it, the horn must have constant negative curvature.[1]

This horn should not to be confused with another popular horn in mathematics, Gabriel's horn or Torricelli's trumpet. The name refers to the religious tradition of identifying the archangel Gabriel as the angel who blows the horn to announce Judgment Day, associating the infinite with the divine. This shape was invented by Italian mathematician Evangelista Torricelli (1608–1647) and therefore sometimes is also called Torricelli's trumpet. Mathematically, it can be described as the surface that results from rotating the graph of $y = 1/x$, for $x \geq 1$, about the x-axis. The result is an interesting mathematical paradox: the resulting object has infinite surface area but finite volume. This is also called the "painter's paradox"—to paint the

Viennese or French horn
based on the pseudosphere.

Gabriel's horn
or Torricelli's trumpet.

Moon flower.

surface, one would need an infinite amount of paint, but to fill the object with paint, one would only need a finite amount of paint.

It is also possible to see the tractrix and pseudosphere in nature. For example, when a moon flower opens at night, its shape is like a pseudosphere.

One of the theoretical models in cosmology used to describe the shape of the universe is called Picard's horn, after French mathematician and physicist Emile Picard (1856–1941), who described the upper half-plane model in hyperbolic geometry as a pseudosphere in 1884. The Picard horn model was an attempt to describe geometric properties of microwave background radiation.[2]

Sometimes the pseudosphere shape is used in buildings and in lamps.

Crocheting a Pseudosphere

To precisely crochet the pseudosphere, we would have to start from a point, but that is not physically possible. Therefore, you should start with a tiny circle.

I suggest you start with three or four chain stitches, join them together, and then continue on a spiral in the same way as for the hyperbolic plane model—by choosing a ratio you will stick with throughout the whole project.

The other way to start is by making a large loop with your yarn, leaving some yarn trailing from this loop. Then, with the hook, draw the yarn through this loop. This initiates an adjustable ring. Now work six stitches in this ring. Once this first round is finished, pull the hanging tail tightly and close the ring.

Patio roof
in Menton, France.

A lamp in the shape of a pseudosphere from Morocco, spotted in the Diaspora store in Ithaca, NY.

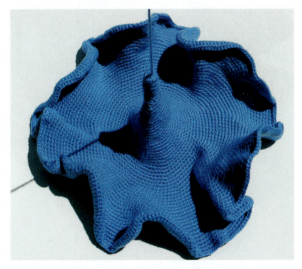

Crocheted pseudosphere with ratio 5:6
(increase in every sixth stitch).

Now you can start to crochet the pseudosphere with the ratio you choose for it. You will notice that at first your project will form like a cone, and then it will flare out and begin to form ruffles. If you were to crochet far enough, then the ruffles would eventually absorb this conical shape.

Crocheting a Symmetric Hyperbolic Plane

We talked in Chapter 1 about a crocheted hyperbolic plane being a nice physical model that cannot be described by analytic equations. In order to crochet a symmetric hyperbolic plane, we first have to do some calculations to make sure that the surface we are crocheting is a surface with constant negative curvature.[3]

Even using the formula that allows you to calculate how to increase stitches in your work, be aware that there is no general recipe, because the calculations depend not only upon the yarn you are using and upon the radius you want to have for your hyperbolic plane, but also—and most of all—upon the tightness of your work and your

Crocheted pseudosphere with ratio 11:12
(increase in every 11th stitch).

Symmetric hyperbolic plane.

crochet style. Below we will work out an example of how you should do calculations for your own work.

We will be using the formula

$$C = 2\pi R \cdot \sinh(r/R),$$

or in different form

$$C = \pi R \cdot (e^{r/R} - e^{-r/R}),$$

where

- C is the intrinsic circumference of a circle with intrinsic radius r on a hyperbolic plane with a radius R (*intrinsic* means that this distance is measured along the surface of the hyperbolic plane; we have talked about intrinsic measurements that Gauss did on surfaces and will discuss intrinsic properties more in Chapter 7);

- sinh is the hyperbolic sine function, which can be found on a graphing calculator;[4]

- R is the radius of the hyperbolic plane you want to crochet (see Chapter 2);

- r is the intrinsic radius of a circle (a symmetric hyperbolic plane will consist of crocheting "concentric" intrinsic circles).

The radius r depends on the height of a crocheted stitch. This is the place where your individual style and the yarn you are using come into the calculations. You should crochet about 12 short rows (10 stitches in a row is enough) with single crochet and measure the height of the middle ten rows of your crochet piece. Then divide that measurement by 10, and that will be the average height of your crochet stitch.

Let us calculate an example where the height of the average crochet stitch is $h = 0.55$ cm and the radius of the finished hyperbolic plane will be 3.5 cm. Then, the intrinsic radius of the nth row is $r_n = n \cdot h$. Let us now use the above formula to calculate for each row the intrinsic circumference $C(n)$,

$$C(n) = \pi R \cdot (e^{n \cdot h/R} - e^{-n \cdot h/R}),$$

and the ratio $C(n)/C(n-1)$, which determines how to increase stitches, where n is the number of the row ($n = 1, 2, 3, 4, \ldots$). For crocheting, we need to find places in our table where the ratio is of the form of a fraction $(k+1)/k$.

n	$C(n)$	$C(n)/$ $C(n-1)$	Nearby fraction	Increase ratio
1	3.5			
2	7.0	2.0	2/1	1 to 2
3	10.8	1.5	3/2	2 to 3
4	14.8	1.4		
5	19.1	1.3	4/3	3 to 4
6	23.9	1.25	5/4	4 to 5
7	29.4	1.23		
8	35.5	1.21		
9	42.6	1.20	6/5	5 to 6
10	50.6	1.19		
11	60.0	1.19		
12	70.8	1.18		
13	83.4	1.18		
14	98.0	1.18		
15	115.1	1.17	7/6	6 to 7
16	135.0	1.17		
17	158.2	1.17		
18	185.4	1.17		
19	217.2	1.17		

The second column is the length of the row in centimeters: the number of stitches will depend on the width of your stitch.

For the yarn used in this example (Red Heart, acrylic, four ply), the stitches were about 0.55 cm wide (the same as the height), and the project was started with three chain stitches joined in a circle, then one chain stitch to start the first row, and then in the first row two single crochets in each chain resulting in six stitches that make up the approximate 3.5 cm circumference of the first row.

You can continue to crochet in a spiral, but you should mark the place where you start a new row. In the second row, crochet two stitches in every stitch of the first row. In the third row, the increases are in every other stitch; the same goes for the fourth row. Then, in the fifth row, crochet three stitches and increase in the fourth stitch. In the sixth row, crochet four stitches and increase in the fifth; the same goes for the seventh and eighth rows. Then, the increase changes in the ninth row to crocheting five stitches and increasing in the sixth one. Continue this ratio for rows 10–14. Then, the ratio changes in the 15th row—crochet six stitches and increase in the seventh. As we can see from the table above, after that the increase ratio stabilizes, so for further rows you just keep the ratio of 6 to 7.

This is a nice property of the symmetric hyperbolic plane: once you go out far enough, the increase stays the same. The crucial part is to find where this "far enough" is. Why is this happening? It is easier to see from the formula in its exponential version. While r is small, the value of $e^{-r/R}$ is significant, but once r becomes big enough, then the value of $e^{-r/R}$ becomes closer and closer to zero and the formula becomes

$$C(n) = \pi R \cdot e^{n \cdot h/R},$$

for large n, which clearly shows exponential growth for the number of stitches.

Let me repeat it again: these calculations are just an example. You have to do calculations for yourself to de-

Another symmetric hyperbolic plane.

termine in which rows you will be changing the ratio of the increase. It will be different for different yarns and for different people because each individual has his or her own crochet style, just as each of us has our own handwriting.

How Big a Hyperbolic Plane Can We Crochet?

Look at the following picture and try to guess which of the four shades of purple color covers the most area.

What was your guess? Each of these shades was measured to be exactly 100 m of yarn—thus the areas of the four colors are all the same! But, as you can see in the picture, the fourth shade was only enough to crochet a couple of rows.

Symmetric hyperbolic plane from four shades of purple color. Which has the largest area?

The pink symmetric hyperbolic plane in the next picture is crocheted from 88 skeins of Moda Dea Ticker Tape ribbon. Each skein weighs 50 g, so the weight of this plane is 4.4 kg. The length of each skein is approximately 62 m, and thus more than 5 km of ribbon was used to make it. There are approximately 54 rows of crocheting, and the distance from the center to the edge is 38 cm. The outside circumference (outside edge) is 30.6 m. The total surface area of this model is approximately 3.7 square meters. It took me eight months to complete this model; of course, I had some other things to do in my life besides crocheting...

On the Euclidean plane, the circumference of a circle is about 6.28 times the radius. However, on a hyperbolic plane, the circumference of a circle is larger than the circle with the same intrinsic radius on the plane. For example, let us compare the circumferences of circles with various intrinsic radii on a hyperbolic plane with radius 20 cm.

Radius of a circle	Circumference on a plane	Circumference on a hyperbolic plane
10 cm	62.8 cm	65.4 cm
1 m	31.4 m	100 m
5 m	157 m	49,000,000 km

Forty-nine million kilometers is more than 100 times the distance from the Earth to the moon!

Global warming (symmetric hyperbolic plane, 2008).

Chapter 7

Metamorphoses of the Hyperbolic Plane

Endless Possibilities for Play

Crocheted models of the hyperbolic plane are very inviting to play with. There are many different ways to fold symmetric hyperbolic planes. Starting from the same basic plane, there are unlimited possibilities for creating different fiber sculptures.

Do these different shapes have anything to do with mathematics, or are they purely aesthetic forms? From the pictures, each set of shapes in the same yarn all started with the same basic crocheted model; the rest is just shaping them differently. What do these changing shapes have to do with the mathematics of hyperbolic planes?

Nobody in my family ever was a mathematician. When I first started to crochet models of the hyperbolic plane, one of my family members frowned that no serious mathematician would waste time on such nonsense. It says something about the general misconception of how mathematicians think and how they work. In movies, most of the time, if there is a mathematician depicted doing research, then he (extremely rarely, she) is buried in papers and books or is writing thousands of fancy formulas on a blackboard.

George Orwell once suggested watching how a waiter enters a dining hall:

> It is an instructive sight to see a waiter going into a hotel dining-room. As he passes the door a sudden change comes over him. The set of his shoulders alters; all the dirt and hurry and irritation have dropped off in an instant. He glides over the carpet, with a solemn, priest-like air... he entered the dining-room and sailed across it, dish in hand, graceful as a swan.[1]

What does this quote tell us about mathematicians and how we explain mathematics to a general audience? Let us think, what is the purpose of separating the front

Endless possibilities of shaping the hyperbolic plane.

(fancy dining room) from the back (busy, sometimes quite messy kitchen)? It is not only to keep customers from interfering with the cooking. It is also to keep them from knowing too much about the preparation process: to the customer, the completed dishes appear "magically" before them. The front and the back of mathematics are not physical locations like the dining room and the kitchen. The front is mathematics in its finished form: lectures, textbooks, journal articles. The back is mathematics among working mathematicians.[2] William Byers wrote in his book *How Mathematicians Think*:

The most pervasive myth about mathematics is that the logical structure of mathematics is definitive—that logic captures the essence of the subject. This is the fallback position of many mathematicians when they are asked to justify what it is that they do: "I just prove theorems." That is, when pressed, many mathematicians retreat back to a formalist position. However, most practicing mathematicians are not formalists—what they really want is usually not some collection of answers—what they want is *understanding*.[3]

With rare exceptions, mathematicians do not get their research ideas from sudden illumination. In most cases, it

Fiber sculptures with constant negative curvature. They all started from the same shape (lower right), were divided into two, three, and four equal parts, respectively, and then were folded.

happens after observing some phenomenon, looking for patterns, then trying to generalize by asking questions like, "What is going on here?" It is like playing with these soft crocheted models, folding them, rolling them, comparing their sizes and then asking, "In what ways are these shapes the same? In what ways are they different?" One way to find answers to these questions is to use the notion of geometric manifolds.[4] If you have never heard about it, it might sound too abstract a notion for you, but let us start with something very simple—let us look first at curves.

Curves: Geometric 1-Manifolds

Manifolds come to us in nature and in mathematics by many different routes. Very frequently, they come naturally equipped with some special pattern or structure, and to understand the manifold we need "to see" the pattern. At other times, a manifold may come to us naked; by finding structures that fit it, we can gain new insight, relate it to other manifolds, and take better care of it.

—W. Thurston[5]

Sometimes this understanding, which Byers talks about, comes from the ability to find a different way of looking at something familiar. Riemann, in his inaugural address in 1854, was the first to introduce the notion of a manifold as a mathematical approach to exploring different regions of space. The name *manifold* comes from Riemann's original German term, *Mannigfaltigkeit*, which William Kingdon Clifford translated into English as "manifoldness." The simplest example of a manifold is a line; that is, we draw a line that we imagine going off forever in either direction. To understand manifolds, let us start with a piece of string and a circle drawn on a piece of paper.

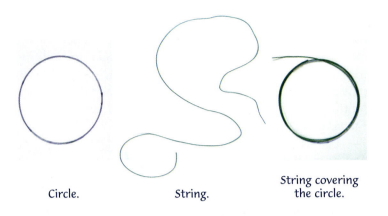

Circle.   String.   String covering the circle.

Now lay the string on the paper so that it covers the circle (and only the circle)—you can do this by starting with one end of the string and laying it along the circle in one direction, allowing the string to overlap itself as necessary.

The string represents a line that we can imagine goes on and on forever in both directions (mathematicians would call this property "completeness"). We can imagine the circle as a curve without thickness; it is complete because you can go along it in either direction without reaching an end. In what ways are a line and a circle the same, and in what ways are they different? There are a lot of possible answers to these questions: for example, both the line and the circle are curves, but you would say the line is straight and the circle is round. The properties of being "straight" or "round" are properties that depend on how the line and the circle are placed in their surrounding space.

What we are interested in here is what the line and the circle are like in themselves and not in relationship to their environment. Mathematicians express this by saying that "straight" and "round" are *extrinsic* properties; we are interested in *intrinsic* properties that depend only on observations and measurements made along the curves. In-

trinsically, on the line or the circle, if you start at any point you can move any distance in either of only two directions. This completely describes the *local intrinsic geometry* of the circle and the line. Mathematicians say that the line and circle are *locally isometric* ("isometric" means "same measurement").

The intrinsic difference is that on the string (line), you never get back to where you started, but on the circle, if you travel in either direction some fixed distance C, you will return to where you started. Mathematicians say that the line is *simply connected* and that the circle is *not simply connected*; more intuitively, the circle encloses a hole and the line does not.

In mathematics, both the circle and the line are called a *geometric 1-manifold*, or simply a curve. We can say that the line *covers* the circle; and, in that sense, the line is the base for geometric 1-manifolds. Any closed curve drawn on paper is a geometric 1-manifold and can be covered by a line. The closed curve is *intrinsically* the same as the circle with the same perimeter (circumference) in the sense that all measurements along the two curves are the same; mathematicians say that the circle and the closed curve are the same geometric 1-manifold or that they are *isometric*.

Circle and closed curve from the same loop of string.
They are isometric—the same geometric 1-manifold.

Flat Geometric 2-Manifolds

Now let us do another experiment: Take a sheet of paper and lay it on a table. You can imagine that it extends in all directions to become a piece of the so-called Euclidean plane. Now press together opposite sides of the sheet of paper as in the picture: the result is a cylindrical surface. You can imagine this surface extended forever in the two "open" directions. We can ask the same question as before: in what ways are the plane and cylinder the same and in what ways are they different?

Mathematicians say that the plane *covers* the cylindrical surface and that this cylindrical surface *is locally iso-metric* to the Euclidean plane—similar to the relationship we just saw between the circle and the line. A small region on the cylinder (that does not go all the way around) intrinsically is the same as a region on the plane in the sense that all measurements on the surface will be the same as on the plane. The plane and the cylinder are called *flat* (or *Euclidean*) *2-manifolds*.

Let us continue our explorations: Unroll the paper and draw a straight line on this sheet of paper. Now roll this paper again into a cylinder. The line is still intrinsically straight. Draw some other straight lines on the unrolled sheet of paper and then roll it again. What kind of intrinsic straight lines can you find on your cylinder?

By now you have noticed that when the plane covers (is wrapped around) the cylinder, the curves on the cylinder that are covered by straight lines on the plane are intrinsically straight on the cylinder. (In mathematics these

Plane and cylinder are locally isometric.

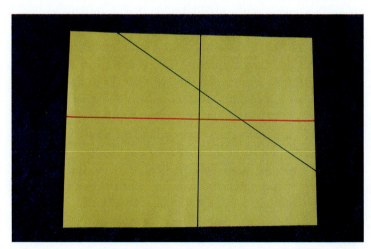

Lines on a piece of paper that is rolled into a cylinder.

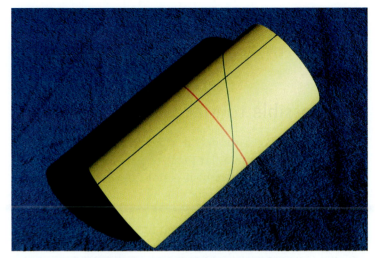

Each color shows one of the three kinds
of intrinsically straight lines on a cylinder.

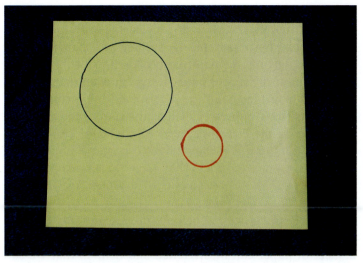

Circles on a plane can be shrunk to a point.

intrinsically straight lines are called *geodesics*.) The intrinsically straight lines on the cylinder are of three kinds (see picture): circles around, lines along, and helices.

For each point on a cylinder, the maximum curvature of a curve through this point is $1/R$ (for a circle around the cylinder) and the minimum curvature is zero (for a straight line along the cylinder). The curvature of the cylinder is a product of these two perpendicular curves and is therefore zero. This is another reason to call the cylinder a flat surface.

However, the cylinder is different from the plane in that you can draw on the cylindrical surface a circle (one that goes around the cylinder) that cannot be shrunk *on the surface*. Mathematicians say that the plane is *simply connected*, because it has a property that every closed curve can be shrunk on the surface, while the cylinder is *not simply connected*.

However, on a cylinder, not all circles can be shrunk.

All curves are geometric 1-manifolds, but this is not true of different surfaces: not all surfaces are geometric 2-manifolds. For example, the surface of a banana or an egg is not a geometric manifold because the local geometry is different at different points. Gauss showed that the curvature of a surface could be determined by intrinsic measurements (see Chapter 2) such as the sum of the angles of a triangle or the circumferences of circles of various (intrinsic) radii. So this means that two portions of a surface cannot be isometric unless they have the same curvature. This tells us that *only surfaces with constant curvature can be geometric 2-manifolds*.

There are other flat geometric 2-manifolds covered by the plane: imagine that the cylinder you made out of a sheet of paper can be stretched and bent so that we can glue the two open ends together. The surface would look like the surface of a bagel or a doughnut. This surface, shown below, is called a *torus*.

However, the torus is not flat. It is *not* a geometric 2-manifold because it does not have constant curvature and because the geometry of the plane that covers it has been distorted.

Is It Possible to Make a Flat Torus?

One cannot make a model of a flat torus in our three-dimensional space from a flat piece of paper without distorting the paper. At this point, it is not important whether or not you can physically make this model; you certainly can imagine it: glue opposite edges of a rectangular piece of paper to form a cylinder and then imagine gluing opposite ends of this cylinder to each other (you could do this without distortion if you could be in 4-space). The flat torus is a geometric 2-manifold.

There is another way to "make" the flat torus. Imagine that instead of a rectangular shape, you start with a regular hexagon.

Cylinder distorted into a (not flat) torus.

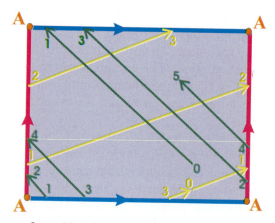

Flat torus formed by gluing together the same-colored arrows. Green and yellow paths are each a single straight line on the torus.

Notice that the hexagon has three pairs of parallel sides. Let us call them opposite sides. Gluing these opposite sides together, we can form the so-called *hexagonal torus* that is also a geometric 2-manifold.

These two descriptions of the torus are closely related to two patterns of tilings in the plane: it is possible to tile the whole plane with identical squares or with identical hexagons.

A flat torus can also be imagined using motion on a computer screen. It was used in early video games. A blip on the screen representing a ball or a rocket traveled in a straight line until it hit an edge of the screen. Then the blip reappeared, traveling parallel to its original direction from a point at the same position on the opposite edge. Check that this is a motion that would happen on the surface of a flat torus. Try playing some games on a flat torus that are

available on Jeff Weeks' webpage, www.geometrygames. org.

Now imagine the same screen, but let us set different rules for the motion. The ball will travel again in a straight line until it hits an edge of the screen. But now its "return" motion will depend on which edge it hits. When it hits the top or bottom edge, then the blip will proceed exactly the same as for the flat torus (traveling parallel to its original direction from a point at the same position on the opposite edge). However, when the blip hits a vertical edge of the screen, it reappears on the opposite edge but in the diametrically opposite position and travels in a direction with slope that is the negative of the original slope. You can check that following these rules, one can continue this motion forever. This is the other example of a flat 2-manifold, the so-called *flat Klein bottle*. We will return to it a little bit later.

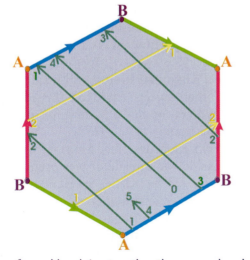

Flat torus formed by gluing together the same-colored arrows. Dark green and yellow paths are each a single straight line on the torus.

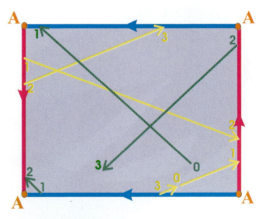

Flat Klein bottle formed by gluing together the same-colored arrows. Green and yellow paths are each a single straight line on the flat Klein bottle.

Before we do so, let us explore one of the most popular fun objects in geometry. In 1858, German mathematicians August Ferdinand Möbius (1790–1868) and Johann Benedict Listing (1808–1882) independently discovered a one-sided surface that is now called a *Möbius strip* (or *band*). If you are not familiar with it, you should construct one from a (preferably long) strip of paper. Glue the ends of a strip together but before you do so, make a half twist.

To begin your explorations, draw a line down the middle of the Möbius strip, parallel to the edges, and then cut along this line. Before you finish cutting, try to guess whether you will get two pieces or just one. On another Möbius strip, mark a line starting about one third of the way from the edge and then cut parallel to the edges. This

cut will give you a different result. You can experiment with making other Möbius strips with more than one half twist before gluing the paper strip together.

Another experiment you can do is to make a Möbius strip out of *clear* plastic and draw on a separate small piece of *clear* plastic a face in profile. Move the face along a line that you draw in the middle of the clear Möbius strip. What happens to the face when you come back to where you started? These properties of the Möbius strip you just experienced have been important in understanding different geometries on various surfaces.

In mathematics, the Möbius band is an example of a nonorientable surface. A surface is said to be *nonorientable* if a shape drawn on it can be transformed into its mirror image simply by moving the shape along a path on a surface. (Otherwise, a surface is *orientable*, and a shape maintains its orientation no matter along what path it is moved.)

The Möbius strip fails to be a geometric manifold only because it has an edge (note that there is only one edge!), and so it is an example of what is called a *geometric manifold with boundary*.

If you are a passionate knitter and want to knit a Möbius scarf or other objects based on this idea, I suggest two books.[6] Quilters can find a pattern for quilted Möbius bands in another book.[7]

In 1882, Felix Klein described a surface with no distinction between inside and outside. It was originally called in German *Kleinsche Fläche* (Klein's Surface); however, it was incorrectly interpreted as *Kleinsche Flasche* (called Klein bottle in English), which ultimately led to the adoption of this term in the German language as well. In his original paper, Klein introduced it as a certain surface that can be visualized by inverting a piece of a rubber tube and by letting it pass through itself so that inside and outside meet. If

Möbius strip made from a strip of paper.

we try to realize a Klein bottle in three dimensions, we get a self-intersecting surface. The true Klein bottle can exist in four dimensions, and there it is not self-intersecting.

However, it is fun to make Klein bottle interpretations in three dimensions. If you have two mirror-image crocheted Möbius strips and sew them together, you get a Klein bottle! Recall from earlier that the Klein bottle is a flat geometric 2-manifold.

The Klein bottle has become one of the favorite objects for mathematical knitters, so you can find many patterns for it on the Internet. Most of these Klein bottles are not geometric 2-manifolds because they distort the flat surface and do not have constant curvature.

Jeff Weeks has designed several computer games on the flat torus and Klein bottle. Clifford Pickover has a very nice collection of facts about the Möbius band.[8]

Felix Klein was very interested in different mathematical models. When he became a professor in Leipzig in 1880, he immediately started to acquire mathematical models (made of plaster, metal, string, and other materials) and to establish a model collection that he used in his geometry lectures. When he moved to Göttingen, Klein, with his colleague Hermann Amandus Schwarz, expanded his new department's collection of mathematical models and instruments up to 500 models. Following the success of these mathematical models, in 1893 the Prussian government decided to participate in the World's Columbian Exhibition in Chicago with a university exhibition. Klein and his former student Walther von Dyck organized an Exhibition of Mathematics, including about 100 physical mathematical models and instruments.

Two crocheted Möbius strips, flat geometric manifolds with boundary.

Crocheted Klein bottle: the two different colors are the two Möbius strips sewn together to form the Klein bottle (this is not a geometric manifold because of the distortion).

Glass Klein bottles by Alan Bennett.

Let us try another experiment. It is not a secret that Lewis Carroll was the Reverend Charles Lutwidge Dodgson (1832–1898), better known for his literary works where he tied in some mathematics since he was also a mathematician and logician. In his novel *Sylvie and Bruno Concluded*, Mein Herr teaches Lady Muriel how to construct the "Purse of Fortunatus" by sewing together three handkerchiefs as indicated in this figure.

He justifies the name of the purse as follows: "Whatever is inside that purse is outside it; and whatever is outside it is inside it. So you have all the wealth of the world in that little purse!" Lady Muriel sews together two of the handkerchiefs and examines how she is to sew the third handkerchief along the remaining four edges of the first two. She states, "But it will take time. I'll sew it up after tea." In fact, this is a mathematical joke! Carroll knew that it is impossible to make this surface in Euclidean three-dimensional space. Here it is—another example of the surface that cannot be fully embedded in 3-space! However, you can make your own version of the Purse of Fortunatus.[9]

Purse of Fortunatus sewing scheme:
join the corners marked by the same letter.
(Note that sewing any two together forms a Möbius band.)

Flat 2-orbifold surfaces.

Spherical 2-orbifold:
American football, similar to a rugby ball.

Sewing a Purse of Fortunatus.

The Purse of Fortunatus is not a geometric manifold because each corner of one of the handkerchiefs will be sewn together with only two other corners and thus the total angles around that corner in the completed purse will be $3 \times 90° = 270°$ instead of the 360° for a point in the plane. However, the Purse of Fortunatus is a geometric manifold at all other points. Mathematicians call such a surface a flat (or Euclidean) 2-*orbifold*. Other such flat 2-orbifolds are cones or the surfaces of any polyhedron— these surfaces are flat geometric 2-manifolds except at the vertices (corner points).

One can also have spherical 2-orbifolds, which are spherical (constant positive curvature) geometric 2-manifolds except at the two tips.

We will talk about the applications of orbifolds to music in Chapter 9. More about orbifolds can be found in the book *The Symmetries of Things*.[10]

Hyperbolic Geometric 2-Manifolds

Look again at the pictures in the beginning of this chapter. Now you can hopefully conclude that all those surfaces are geometric 2-manifolds. Each of them is covered by the hyperbolic plane, and so each is locally isometric to the hyperbolic plane (and to each other). However, among these geometric 2-manifolds, only the hyperbolic plane is simply connected—all the other hyperbolic surfaces have holes or circles that cannot be shrunk on the surface.

We can deliberately construct other hyperbolic 2-manifolds with crochet. Also, surfaces can be cut and glued together to create another surface. For example, take two bagels, slice a disc off of each one, and then place them together along those slices, smoothing off the corners. The result is a two-holed bagel. Mathematicians call the surface of this two-holed bagel a *two-holed torus*.

Two-holed torus or surface of a two-holed bagel: this shape sometimes is called an anchor ring.

Will this new surface be a geometric 2-manifold? Look at the points A and B in the picture: the surface is positively curved at point A and negatively curved at point B, so intrinsic geometry near these points will be different. If, for example, we draw a small triangle around point A, then the sum of the interior angles of this triangle will be greater than 180°, but the sum of the angles of a similar triangle around point B will be less than 180°. Thus, the surface of the two-holed bagel does not have constant curvature and cannot be a geometric 2-manifold.

So now let's try to make the two-holed torus into a geometric manifold. The first step is to cut a two-holed torus into a (very irregular) octagon.

This octagon will be far from a regular octagon because its sides will have different lengths and, even worse, it will be "bumpy" because it has both positive and negative curvatures. We then flatten and shape this octagon into a regular plane octagon, keeping track of which edges must be glued to which other edges.

If we glue the edges of this regular octagon, then we would get a surface that is a flat 2-manifold except at the one vertex, where all the vertices of the octagon meet. At that vertex the angle would be eight times the interior angle of the octagon, which is clearly much more than 360°.

Can we make a two-holed torus that *is* a geometric manifold? Let us recall the connection between tilings in the Euclidean plane and geometric manifolds: we could use a square or a hexagon to make a geometric 2-manifold because with these shapes we can tile the Euclidean plane. We now found that we can make a two-holed torus from a regular octagon.

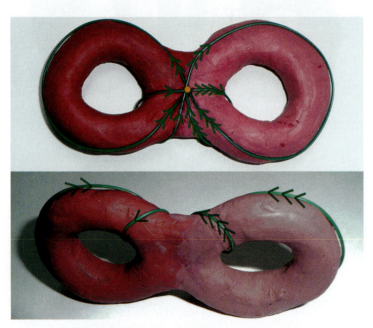

Two-holed torus surface cut into an octagon:
the green lines indicate the cuts.

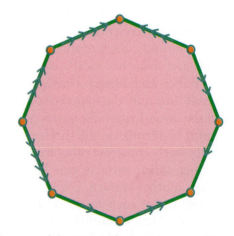

Diagram of the octagon's identifications (gluings)
after the cutting.

Can we tile the Euclidean plane with regular octagons? In the Euclidean plane, the interior angles of a regular octagon each equal 135°.

To tile the plane, we have to put vertices together so that there are no overlaps or gaps, *and* so that at each vertex the sum of the angles comes to 360°. Two octagons would give us 270°, but three octagons would give 405°. Thus, we cannot tile the Euclidean plane with regular octagons, and we cannot use a flat octagon to make a flat geometric 2-manifold.

But what happens if we do not stick with the Euclidean plane? To tile the plane with squares we need four squares with 360° ÷ 4 = 90° at each vertex. To tile the plane with hexagons, we need 360° ÷ 3 = 120° at each vertex. If we want to use the above gluing of a regular octagon, we would need angles of 360° ÷ 8 = 45°—each interior angle of the octagon would have to be 45°. Such a regular octagon does not exist in the Euclidean plane! The angle 45° is an exterior angle of the regular octagon, not an interior angle.

But we have our hyperbolic plane model available. Can we tile it with such octagons? First, to see that such an octagon (with 45° angles) exists in the hyperbolic plane, imagine placing a small (regular) octagon on the hyperbolic plane. Because the octagon is small, its interior angles must be very close to the interior angles (135°) of an octagon in the Euclidean plane. Now let the small octagon grow, keeping it always regular. As happened with triangles (remember Chapter 2), the interior angles of the octagon will continuously decrease in size until, as the vertices get farther and farther away, the angles will become near zero. Somewhere in this process there must be a step at which the interior angles of the regular octagon are 45°. We will now show how to crochet this octagon.

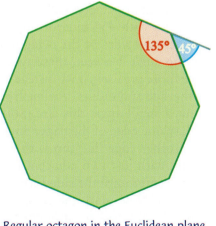

Regular octagon in the Euclidean plane and its interior and exterior angles.

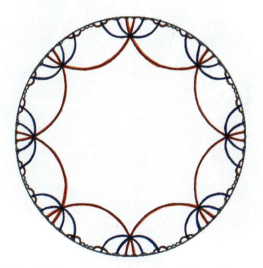

Tiling the hyperbolic plane with regular octagons in the Poincaré disk model.

Crocheting the Hyperbolic Octagon

One of the very first models of the hyperbolic plane I made in 1997 was a "hyperbolic pair of pants." Theoretically this model was known to mathematicians.[11] But it was never physically made to touch. Below is the description of how to make a hyperbolic pair of pants that is a geometric 2-manifold. You can imagine an octagon cut out in the shape of a stop sign from a stretchy material and then every other side glued or sewn together. Visually the result may look like a pair of pants, but in some places it will be stretched more than in others and therefore won't be a geometric manifold.

Regular octagon on a hyperbolic plane with 45° interior angles.

Let us first make a regular octagon on a hyperbolic plane model. First we need to divide the center angle into eight equal parts (45° each). This can be done by folding the same way as you would a piece of paper. Fold first to get a straight line. Mark it with stitches. Next fold the perpendicular to the first line, and mark with stitches again. Now you have divided the center angle into four equal parts. Dividing each of these angles in half will result in the center angle being divided into eight angles, each of which is 45°.

The picture features the hyperbolic plane model (with ratio 12 to 13) with an octagon whose interior angles are 45° each.

There is really no precise procedure for constructing the 45° octagon we need. After you construct eight 45° angles around the center, mark equal distances in all eight directions (in the model shown in the picture, the distance is approximately 15 cm). Construct one of the sides of the octagon by folding the line between the marks on two adjacent lines and marking it with stitches.

Notice that any distance will give you a regular octagon, but not every regular octagon will give us the required model: we need one with 45° interior angles. Therefore, after you construct two sides of the octagon, you have to check whether the angle between these two sides is 45°. This can be done more precisely if we instead check that the angle between the side and radius (the line from the center) is 22.5°. Cut out a small angle this size (360°/16) from a piece of paper. Lay one edge of the paper angle along a ray and see how the other edge lines up with the side of the octagon. Working with just two adjacent rays, adjust the distance from the center until the side forms the necessary angles with the rays. Now mark the equal distances from the center along all eight rays accordingly. Once you have the octagon constructed on a piece of hyperbolic plane you have crocheted, you will have a pattern for crocheting a hyperbolic octagon.

It is important that you crochet your own pattern because for different people both the tightness of crocheting and the materials used will vary. Start with a chain, then crochet according to the instructions for making a simple hyperbolic plane, following the same ratio of increase you had for the model on which you did the previous construction. When the intrinsic radius of the row gets longer than needed for your octagon (i.e., than the distance you measured in the previous construction), then simply stop crocheting that row any longer. Do not worry about the sides not being too neat—they will be folded inside in the final model. If you want your model to be more symmetric, you can start with a symmetric hyperbolic plane as in the preceding figure.

When your octagon is done, attach pieces of Velcro (fabric hook-and-loop fastener) to the sides in the following way: first attach a "hooks" piece along one side of the octagon, leaving the next side empty, then attach a "loops" piece to the third side, leaving the fourth side empty. Continue this pattern: "hooks," empty, "loops," empty.

When you stick one "hooks" side to one "loops" side and stick the remaining nonempty sides together, the shape will remind you of a pair of pants as you can see in the picture below. That is why this model is also called by mathematicians the "hyperbolic pair of pants."

Symmetric plane with octagon construction on it.

Hyperbolic pair of pants from an octagon.

Hyperbolic pair of pants with hook and eye.

It is interesting to calculate the area of such hyperbolic regular octagons with 45° interior angles. The area of a hyperbolic triangle[12] can be found as

Area of hyperbolic triangle =
(180° − sum of the angles of the triangle)
× (π/180°) × (radius of the hyperbolic plane).²

According to our construction, the octagon consists of eight hyperbolic triangles with interior angles 45°, 22.5°, and 22.5°. The area of each triangle is

$$(180° − (45° + 22.5° + 22.5°)) \times (\pi/180°) \times R^2 = (\pi/2)\,R^2,$$

and the area of all eight triangles is $8 \times (\pi/2)\,R^2$ or $4\pi R^2$, where R is the radius of the hyperbolic plane. You might remember from high school geometry that in Euclidean geometry, this is the formula for the area of the surface of a sphere with radius R.

Two-Holed Torus from Four Right-Angle Hexagons

There are other ways to make geometric 2-manifolds. One of the most cited ways in the mathematical literature of making a pair of hyperbolic pants is to "glue" together four right-angle hyperbolic hexagons with equal sides.[13] You can construct right angles by folding as follows:

1. First mark one straight line, then fold it "in half" to get a perpendicular line, which you mark.

2. Choose a point on the first line and fold to mark another perpendicular (see picture on the next page).

3. Now choose points on both perpendiculars that are the same distance from the original line as the perpendiculars are from each other.

Once you attach the sides of the octagon, you now have two smaller openings—the circumference of each is a length of one side of the octagon—and one larger opening—the circumference of which is made of two sides of the octagon. Notice that you have two vertices left to attach to each other—this can be done with a single dressmaker's hook and eye as in the picture above.

Now there are four openings of the same size. We would need to be in the fourth dimension to be able to glue these openings pair-wise together. The result would be a two-holed torus that is a geometric 2-manifold and whose intrinsic geometry is hyperbolic. We cannot perform this gluing in three-dimensional space because that would destroy the intrinsic geometry of the surface—in everyday language this means that it would stretch and/or buckle. To preserve the intrinsic geometry, we cannot allow the surface to buckle or stretch, but this gluing can be imagined in 4-space.

4. At each of these new points, fold to mark additional lines perpendicular to the second and third lines.

5. These last two perpendiculars will intersect to form a hexagon. If the hexagon is regular, then you are done. If the hexagon is not regular, then adjust the distance between the first two perpendiculars and try again.

On the Euclidean plane, you can construct regular hexagons of any size, but they will all have the same interior angles. On the hyperbolic plane, you can also construct any size regular hexagons, but their interior angles will differ depending on the length of the sides. Unfortunate-

ly, there is no practical formula to find the length of this regular hexagon because any formula will depend upon the radius of your particular hyperbolic plane—and this is difficult to measure precisely. When making a model, it is best to create geometric constructions by folding and trial and error.

On the hyperbolic plane model, mark two right-angle hexagons and sew Velcro as shown in the picture. There are two ways to "glue" these sides together: make sure you choose the way in which all openings are the same after gluing. Do the same on a second hyperbolic plane so that you have two identical models each made from two right-angle hexagons.

How to start construction of a right-angle polygon.

Attaching Velcro on the two right-angle hexagons.

Four right-angle hexagons, with six equal-sized openings.

Four glued-together right-angle hexagons can produce infinitely many geometric 2-manifolds: you can glue together the middle openings after any rotation, producing a new manifold with each rotation.

The two-holed torus models below cannot be glued together in 3-space without distortion, but we can imagine gluing them in 4-space without distortion.

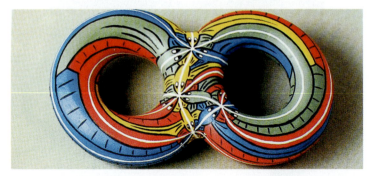

Two-holed torus from four hexagons, not a geometric manifold, by Professor Douglas Dunham, University of Minnesota Duluth.

Other Hyperbolic 2-Manifolds

It is possible to put together four right-angle hexagons in a different way to form a geometric manifold with three holes. These can be glued together in various ways to form more geometric 2-manifolds.

To avoid too much excess material,
I first made this drawing on a big hyperbolic plane model
and then crocheted the necessary shape
using that drawing as a pattern.

Four right-angle hexagons making another surface.

This same surface can be formed in other ways, as in the pictures.

Two views of a two-dimensional manifold formed by wrapping two ideal triangles.

Bill Thurston's paper model, which shows how to get a pair of pants by wrapping ideal triangles. This was given to me as a "crochet assignment."

This model is formed from two ideal triangles, each with three parts (with different colors) and a hole in the middle. The "fluffy ears" show how this surface would continue.

Why Geometric Manifolds?

Why do we talk so much about these different two-dimensional surfaces, or geometric 2-manifolds? A quote from Thurston follows in which he describes some of the special important properties of surfaces—it is not necessary for the reader to understand all terms in order to get the flavor of the importance and power of two-dimensional geometric manifolds:

> Two-dimensional geometry is a special case, in many ways. As a start, there are infinitely many regular polygons. Regular polygons, unlike polyhedra in any higher dimension, are flexible. The group of isometries of the plane is solvable. The geometry of similarity in the plane is essentially the same as complex arithmetic. Topology in two dimensions is also a very special case. The topology of a closed oriented surface is measured by a simple invariant, the Euler characteristic. Every oriented surface is a complex 1-manifold, and in fact, any Riemannian metric on a surface determines a unique conformally equivalent complex structure. The list goes on and on: there are many phenomena that do not readily generalize to higher dimensions. This is a feature, not a bug: because two dimensions is a special case with many special features, two-dimensional topology, geometry and dynamics form an extraordinarily rich, beautiful and unique ecosystem that ends up being highly connected to a large array of other topics in mathematics and science.[14]

We can also have higher-dimensional geometric manifolds. A geometric manifold is a space that is locally isometric on 2-, 3-, or more-dimensional space. You can imagine this like a pattern cut out from a fabric and then patched together seamlessly from small pieces.[15] As an example of this idea for 2-manifolds, see the right-angle hexagons used earlier in this chapter.

Geometric manifolds appear, for example, in mechanics. When engineers are constructing a robot, the space describing the limb postures and locations of a robot is typically described by an abstract geometric manifold.[16]

It is not known yet what the shape of the physical universe we live in is, but it is assumed that the physical universe is a geometric 3-manifold. There are many possible shapes for this 3-manifold. This is a topic of Jeff Weeks' book,[17] which I highly recommend if you are up to exploring what the three-dimensional manifolds look like. More about 3-manifolds is also in Chapter 24 of *Experiencing Geometry*.[18]

Chapter 8

Other Surfaces with Negative Curvature: Catenoid and Helicoid

Saddle Shape and Minimal Surfaces

It is easy to see whether you are dealing with a flat two-dimensional surface: just put this surface on a table and try to flatten it. If it lies evenly, then it is flat or, as you now know, it is a surface with zero curvature. But if the surface is not flat, how can you find out whether it is positively or negatively curved?

In the nineteenth century, Carl Friedrich Gauss realized that it was possible to determine the shape of a curved surface by making exact measurements along its surface. One way to do this is to start at a point and measure the circumference of circles centered on that point with various intrinsic radii (radii measured along the surface). For the Euclidean plane, the circumference of a circle of radius r is $2\pi r$. On surfaces of positive curvature, the circumference of a circle with intrinsic radius r is less than $2\pi r$, while on a surface of negative curvature, the circumference is greater than $2\pi r$ (Chapter 2).

A saddle-shaped surface is negatively curved, but the difference between a saddle surface and the hyperbolic plane is that the hyperbolic plane has the same negative curvature everywhere, while on a saddle surface, the further you get from a saddle point the more flat the surface becomes.

Many saddle points can be seen on this, *Seahorse*, jewelry piece by Benjamin Storch (photo by Bernd Krauskopf).

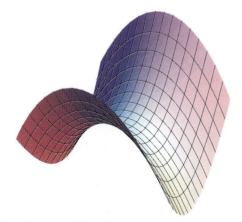

"Saddle-shaped" surface: hyperbolic paraboloid.

Wire frame with soap film.

In this chapter, we will describe a crocheted model of another surface that has negative (but not constant!) curvature. This model is one of the so-called minimal surfaces, the *catenoid*. The term *minimal surface* was coined for surfaces that have some constraints that minimize surface area. For example, if a wire frame is dipped in soap solution, the resulting shape of the soap film solution that sticks to the frame (if it does not enclose a bubble of air) will try to cover the smallest possible area.

The first two minimal surfaces—the catenoid and the helicoid—were discovered in the eighteenth century; the next one was only discovered almost 60 years later, in 1835.

This discovery was regarded as so sensational that its author, German professor Heinrich Ferdinand Scherk (1798–1885), was awarded a special prize. In the mid-nineteenth century, minimal surfaces became known as soap surfaces after Belgian physicist Joseph Antoine Ferdinand Plateau (1801–1883) did a lot of experiments with soap film. Data from his very accurate measurements later served as verification of theoretical results on minimal surfaces.

In the introduction, I mentioned Finnish mathematician Lars Ahlfors, who in 1936 was awarded one of the first Fields medals. The second Fields medal was awarded the same year to the American mathematician Jesse Douglas (1897–1965) for his work in the theory of minimal surfaces. In the 1960s, another mathematician I mentioned in the introduction, Robert Osserman, extensively studied minimal surfaces.[1] Research on minimal surfaces has become of great interest in the past 15 years because of applications in molecular engineering, materials science, and nanotechnology. Atoms in some materials, such as diamonds and starch, lie on structures that essentially are minimal surfaces. The skeletal elements in sea urchins can be described with minimal surfaces, and minimal surfaces are applied in making contact lenses.

The Catenoid

The catenoid is the oldest known minimal surface. It was first described by Leonhard Euler in 1740 and Jean Baptiste Meusnier in 1776. The catenoid is strongly negatively curved near its neck, and the farther from the center you go, the flatter it gets.

A physical model of the catenoid can be made by dipping two circles in any soap solution and then slowly drawing the circles apart—this was shown by Euler in 1744.

Catenoid soap film.

Hanging chain (photo by Laimdota Upeniece).

The catenoid can also be described as a surface of revolution: one can get the catenoid by rotating a catenary curve around a line. The word *catenary* is derived from the Latin word *catena*, meaning "chain." As its Latin origin suggests, a catenary is a very natural curve: it is the resulting shape if you let a cable hang freely with its ends fixed.

Galileo mistakenly thought that such a curve would be a parabola—the shape of a catenary is easy to mistake for a parabola. For example, often the description of the Gateway Arch in Saint Louis, Missouri, is given as an upside-down parabola, but, in fact, that arch follows the form of an inverted catenary. It is 630 feet wide at the base and 630 feet tall. The exact analytic formula

$$y = -127.7 \text{ ft} \times \cosh(x/127.7 \text{ ft}) + 757.7 \text{ ft}$$

is displayed inside the arch.

Catenary-shaped braces are used to support bridges. One such bridge was recently opened on the Cornell University campus in Ithaca, New York.

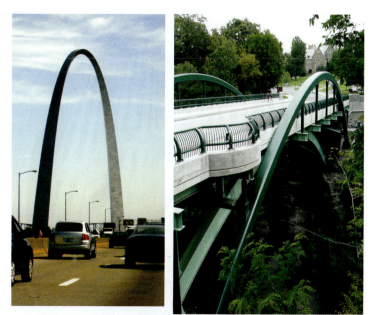

The Gateway Arch, Saint Louis, MO (photo by Beverly West).

Bridge with catenary-shaped support (Cornell University, Ithaca, NY).

The Catalan architect Antoni Gaudí (1852–1926) made extensive use of catenary shapes in most of his work, including the cathedral Sagrada Família. In order to solve for the ideal vault lines, he built inverted scale models of the numerous domes by using hanging ropes to represent stones under compression.

Catenary shapes can also be used as a road cover on which square wheels will roll smoothly. Stan Wagon constructed such a model to demonstrate catenary properties.

Stan Wagon, riding a bicycle with square wheels on a catenary road (Macalaster College, from S. Wagon's personal archive).

Cathedral Sagrada Família, Barcelona
(photo by Lelde Taimina).

The Helicoid

Another well-known minimal surface is the *helicoid*. It was discovered in 1776 by French mathematician, engineer, and army general Jean Baptiste Meusnier (1754–1793).[2] Its name derives from its relationship to the helix: for every point on the helicoid, there is a helix contained in the helicoid that passes through that point. A helicoid is constructed by connecting the two branches of a double helix with horizontal lines passing through the vertical center line. You have already experienced forming a helix by drawing a diagonal line on a piece of paper and then wrapping it into a cylinder. (See Chapter 7.) The line formed is a helix. When you drive up a ramp in a parking garage or walk down a spiral staircase, you are on a helicoid.

Helicoid with the center removed:
a parking garage ramp in Sicily.

DNA double helix sculpture by Charles Jenck
at Clare College, Cambridge.

Double-helicoid auger on a construction site, Cornell University.

Double helicoid with the center removed:
exit stairway in the Vatican Museums.

An auger is a mechanical device for moving material or liquid, using the shape of a helicoid. It is used on construction sites for drilling holes and removing soil. Large augers are used in grain elevators for lifting grains, and fishermen use augers for drilling holes in ice. Augers are also in snow blowers. The earliest use of such tools has been attributed to Archimedes in the third century BC, when he invented the so-called *Archimedean screw* for lifting water. Some augers are derived from a single helix (see Chapter 3) and are thus not (double) helicoids.

Isometry between the Catenoid and Helicoid

Crocheted models allow us to perform physically the famous transformation usually taught in a differential geometry classes: *There exists an isometric transformation that takes the catenoid (cut along one catenary) to the helicoid and vice versa without stretching.*

Follow the steps in the pictures.

Crocheted catenoid and crocheted helicoid.

Start with the helicoid and remove the straight support wires.

Adjust toward a catenoid (cut along one catenary).

Amazingly, the helicoid and catenoid are two minimal surfaces that have the same intrinsic geometry—as, for example, the cylinder and the plane have the same intrinsic geometry (as discussed in Chapter 7).

Crocheting a Helicoid/Catenoid

The model will start from the center line in two directions; in other words, we will crochet one half and then from the same original line crochet the mirror image in the other direction. This first row corresponds to the neck of the catenoid or the height of a helicoid—which are the same. Let us call it Row 0.

Before you start to crochet the model, you will need to crochet a small pattern in order to make some measurements necessary for further calculations. The stitch used in the models pictured above is double crochet, and the yarn is Red Heart acrylic. (Double crochet is performed the same way as the single crochet, except the "yarn over" step is repeated twice.) In your pattern, measure the width and the height of your stitches.

In the model above, Row 0 has length H = 40 stitches. The width of the stitch is w = 0.55 cm, so $H = 40 \cdot (0.55$ cm$) = 22$ cm. The height of the stitch h = 1 cm. Now you need to do some calculations in order to plan where to make necessary increases in your model.

Let us look at the right-angle triangle with legs H and x and hypotenuse L. If we bend the triangle so that leg x turns into the circle with circumference x, then the hypotenuse L will be one turn of a helix on the helicoid. We will start to crochet with Row 0 having length H, and we need to calculate the number of stitches in each consecutive row according to the following formula. For the first row, the height will be equal to the height of the stitch, so we can calculate the length of L. The Pythagorean Theorem gives

$$L^2 = H^2 + (2\pi n \cdot h)^2,$$

so we can calculate length L for the first row as follows:

$$L^2 = 22^2 + (2\pi \cdot 1 \cdot 1)^2 = 484 + 39.44 = 523.44$$
$$L = \sqrt{523.44} = 22.88 \text{ (cm)}.$$

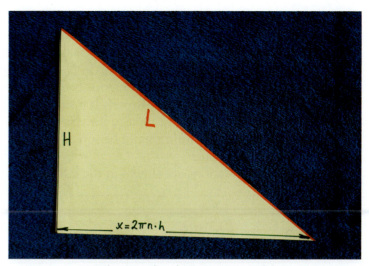

Right triangle relating L, H, x, and h.

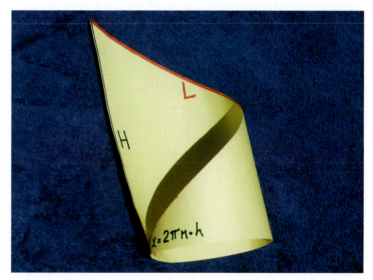

Place leg H of the right triangle perpendicularly to the table, bend the other leg to form a circle on the table, and then the hypotenuse of this triangle will be one turn of a helix.

Now we need to find the number of stitches in this row by dividing the length of the row by the width of the stitch: $L/w = 22.88/0.55 = 41.59$. We cannot make a fraction of a stitch, so we round this number up to $N = 42$. We can now find $\Delta L = 2$, that is, the difference between the number of stitches in Row 0 and in Row 1. It means that if we started with 40 stitches, in the next row we should add two more stitches. This addition should be spread over the whole row, not in one place.

The further calculations are shown in the table.

n	L/w	N	ΔL
0		40	
1	41.59	42	2
2	46.05	46	4
3	52.66	53	7
4	60.7	61	8
5	69.7	70	9
6	79.3	79	9
7	89.4	89	10
8	99.7	100	11
9	110.3	110	10
10	121.1	121	11
11	131.8	132	11
12	142.7	143	11
13	153.7	154	11
14	164.7	165	11

Notice that there is a certain pattern in the numbers for ΔL, except between the eighth and ninth rows where the difference is just due to rounding. The number of stitches to be added at each row grows, but then after some point (here it is Row 10), it stays the same. Geometrically, it means that up to that certain place our surface has negative curvature (not constant!) near the neck, but then it flattens out and the curvature converges toward zero. For making a model of the helicoid, this means that it is enough to crochet just a few rows after ΔL becomes constant—in this model it is 12 rows from the start. The first half of the model is finished.

To complete the second part of the model, return to Row 0 and crochet symmetrically from the other side the same number of rows (in the example it was 12).

Unlike the hyperbolic plane models, this model needs supports. You can use, for example, skewers and flexible wire to support your model as you can see in the pictures at the beginning of the section. Around the edges you should use some wire.

Chapter 9

Who Is Interested in Hyperbolic Geometry Now and How Can It Be Used?

Computer Science

Once I was working on a large model for an upcoming exhibit. One night I had a very strange dream: I was crocheting my model and it was growing larger and larger, and then it was too much to hold in my hands. It got twisted in some very unusual way, and suddenly I realized that if I can get through this twist, I will see the solution of the P versus NP Problem in computer science (see below). As it happens with dreams, I woke up just before the problem was resolved.

I did my graduate work in theoretical computer science, but that was some time ago. I am no longer actively working in computer science and certainly was not trying to solve one of the greatest unsolved problems in computer science. P versus NP is one of seven "Millennium Problems" for solutions of which the Clay Mathematics Institute has offered a million-dollar prize. On the Institute's website, this problem is described as follows:

> Suppose that you are organizing housing accommodations for a group of four hundred university students. Space is limited and only one hundred of the students will receive places in the dormitory. To complicate matters, the Dean has provided you with a list of pairs of incompatible students, and requested that no pair from this list appear in your final choice. This is an example of what computer scientists call an NP-problem, since it is easy to check if a given choice of one hundred students proposed by a coworker is satisfactory (i.e., no pair taken from your coworker's list also appears on the list from the Dean's office), however the task of generating such a list from scratch seems to be so hard as to be completely impractical. Indeed, the total number of ways of choosing one hundred students from the four hundred applicants is greater than the number of atoms in the known universe! Thus no future civilization could ever hope to build a supercomputer capable of solving the problem by brute force; that is, by checking every possible combination of 100 students. However, this apparent difficulty may only reflect the lack of ingenuity of your programmer. In fact, one of the outstanding problems in computer science is determining whether questions exist whose answer can be quickly checked, but which require an impossibly long time to solve by any direct procedure. Problems like the one listed above certainly seem to be of this kind, but so far no one has managed to prove that any of them are so hard as they appear, i.e., that there really is no feasible way to generate an answer with the help of a computer. Stephen Cook and Leonid Levin formulated the P (i.e., easy to find) versus NP (i.e., easy to check) problem independently in 1971.[1]

About a week went by, but the strange dream was still on my mind. It seemed to me so strange to have a connection between crocheting, the hyperbolic plane, and this computer science problem. So, I decided to do some web research, and to my great surprise I came across a paper by French mathematician Maurice Margenstern and Japanese mathematician Kenichi Morita in which the P-NP problem was solved in the special case of a pentagonal grid on the hyperbolic plane![2]

I wrote to Professor Margenstern asking how he came to think about connecting automata (which are abstract machines in computer science) with the hyperbolic plane. He replied:

> I came to hyperbolic geometry now 9 years ago and this was first for fun. In October 1997, I had a discussion with a Japanese colleague (Morita) who is a well-known specialist of reversible cellular automaton. I told him: "you found reversible universal cellular automata in the square grid, in the hexagonal grid and

in the one with the equilateral triangle; this means that you did the job for 3, 4 and 6. Five is missing." He said: "Well, five is impossible." I said: "Yes, in the Euclidean plane it is impossible, but in the hyperbolic plane it is possible, there are the pentagons with right angles which tile the hyperbolic plane." This was the beginning of the adventure. And I was lucky to have discovered very interesting connections between cellular automata and hyperbolic spaces.

Since then there have been more similar publications.[3]

This adventure in computer science prompted me to look at where else the hyperbolic plane could appear and who is interested in it.

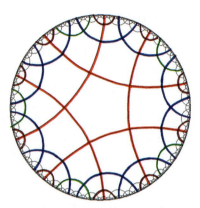

Rectangular pentagonal grid in the Poincaré disk model.

Pentagonal rectangular grid on crocheted hyperbolic plane model.

Mathematics

From Babylonian times, it was known that if squares are built on the legs of any right triangle, then the sum of their areas equals the area of a square built on the hypotenuse. If written as an equation, this property looks like $a^2 + b^2 = c^2$, where a, b, and c are integers. In number theory, equations like this are called *Diophantine equations*. Of course we all know it as the most famous statement in mathematics, the Pythagorean Theorem, though it was known long before Pythagoras by the Babylonians.

Ancient Greek mathematicians became interested in whether solutions can be found to other equations written as $a^n + b^n = c^n$, where a, b, c, and n are integers and $n > 2$. Certainly, most enigmatic was a note found by Fermat's son Samuel on the margins of the copy of Diophantus' Arithmetica that his father had studied: "I have discovered a truly remarkable proof, which this margin is too small to contain." Fermat's proof (that there were no such equations) was never found, and this challenge acquired the name Fermat's Last Theorem. The note was written around 1630, but only in 1995 did the mathematical world congratulate Andrew Wiles with finally proving Fermat's Last Theorem. Of course, he was using mathematical methods from the twentieth century, among them such advanced mathematics as arithmetic elliptic curves and modular functions. It is not the purpose of this book to give insight as to what they are. I just wanted to share my surprise when I learned from Barry Mazur that periodicity on the hyperbolic plane is the setting for the classical theory of modular functions. Connecting modular functions with arithmetic elliptic curves is the way to formulate the conjecture of Shimura, Taniyama, and Weil that implies Fermat's Last Theorem. Here it is—a connection of the hyperbolic plane to number theory![4] As Mazur wrote:

> Number theory swarms with bugs, waiting to bite the tempted flower—lovers who, once bitten, are inspired to excesses of effort!

Something like that can also be said about the hyperbolic plane and hyperbolic space in general. Hyperbolic spaces have found their place in many other fields of mathematics—algebraic geometry, differential geometry, complex variables, and dynamic systems.

Biology

Biologists are interested in hyperbolic geometry because it can describe the way leaves grow. The ruffling of leaves and the folding of fabric are also of interest to physicists and applied mathematicians. Lakshminarayanan Mahadevan, Lola England de Valpine Professor of Applied Mathematics at Harvard University, uses one of my crocheted models of the hyperbolic plane to think about the folding of the brain during embryogenesis, when gray matter packs itself into the tight confines of a skull. In the case of the brain, for example, "you have a flat sheet that is growing rapidly" inside an enclosed structure, the skull. "Brain folding is a difficult problem, which a number of research groups are working on," Mahadevan writes. Rather than tackle it directly, he has been climbing a smaller hill: he has been studying the growth of the seaweed kelp. Kelp, the blades of which can grow as long as 10 meters, lives in shallow ocean waters. Typically, these algal blades are flat in the center and ruffled at the edges. But the amount of ruffling depends on the environment, says Mahadevan, whose interest in the problem was sparked by University of California, Berkeley, biologist Mimi Koehl while Mahadevan was on sabbatical there. If moved into fast-flowing water, the plant loses its ruffles. Transplant the same algal blade back to areas with gentle currents, and the ruffles regrow. Researchers think that the advantage to the plant is to ensure that it wiggles—exposing as much of its surface as possible to the sun, thus maximizing photosynthetic activity—but doesn't flap so much that it breaks in strong currents. "Of course the plant is not a sentient being that knows when to grow or not," Mahadevan explains. "The real question is, how does stress connect to growth?"[5]

Negative curvature around the edges of leaves.

Chemistry

Properties of the hyperbolic plane have been used in chemistry.[6] Some chemists made clever use of the connections between the intrinsic geometry of a surface and how it sits in space. They locally altered the chemistry of an elastic sheet to make it locally expand and contract while changing the temperature of the surface. Instead of molding materials into complex three-dimensional surfaces, they created self-folding materials by changing the spatial concentration of the N-isopropylacrylamide molecule. This molecule undergoes a dramatic, reversible reduction in volume at 33° C. At this temperature, the flat gel disc folds into controllable three-dimensional shapes.

Reinhard Nesper (ETH, Zürich) and Stefano Leoni (Max Planck Insitute) use tilings and patterns on hyperbolic surfaces to do research in structural chemistry.[7]

Medicine

Non-Euclidean geometry has found applications in medicine. For example, Martin Steiner, a computer scientist at the University of North Carolina, uses hyperbolic geometry to develop tools for analyzing brain images. He said:

> When one starts to analyze images, one first tries the simplest tools that are based in linear or Euclidean geometry. But the human brain is too complicated to fit in such space, so often researchers need to go to a nonlinear space and that is when hyperbolic geometry is helpful because with it we can obtain more detailed information. If you just try to apply Euclidean geometry, you step out of the space of valid results.[8]

Geometry has important applications in surgery: the human body is not flat; in different places it has different curvatures. To find ways to perform reconstructive surgery, doctors use computer tomography imaging that analyzes surface geometry, including parts where the surface has negative curvature.

Physics

J. Richard Gott III, Professor of Astrophysics at Princeton University, became interested in a problem of multiply connected universes in 1975. He and his collaborators have developed a method for quantitatively measuring the topology of structures in three dimensions.[9] He has an idea that the universe has a sponge-like structure, where one of the possibilities for this sponge is that it is built from hyperbolic cells, as in the picture. Notice that these cells are like some models in Chapter 7. Hyperbolic models in astrophysics and cosmology were the topic of a special workshop at the Isaac Newton Institute, University of Cambridge, in June 2003.

Cosmic crystallography: hyperbolic models in astrophysics.

Until 1996, cosmologists were giving an experimental preference to the idea of a hyperbolic universe, with negative curvature and in perpetual expansion. The situation has changed with the implementation of new methods for determining the curvature. The precise measurement of the curvature still remains out of the reach of cosmologists, so the question of the shape of our space is still open.[10]

A group of Brazilian physicists is interested in how signals can be transmitted in the hyperbolic plane. This research has led them to propose a new communication system whose advantage is a reduced signal-to-noise ratio.[11]

French physicists J. M. Vigourex and R. Giust were studying multilayers in optics, when they noticed that some of the phenomena in multilayer surfaces can be explained using hyperbolic geometry.[12]

Hyperbolic spaces are also used in ergodic theory and string theory.

Visualizing the Web

The first programs for linking different documents (hypertexts) with hyperlinks were developed in the 1960s. In 1990, the first webpage was created, and computer scientists were actively working on the development of a web browser. Information visualization became a "hot topic" in the early 1990s. Visually impaired people must rely upon their other senses, but other humans process an image and interpret it much faster than reading written text. For example, when looking at a picture, understanding can be almost instantaneous—as in the saying "a picture is worth 1000 words"! If the same situation is given as a sentence, it takes a longer time to process because you need to read the words, interpret their meaning, and then create a visual image in your mind's eye.

One of the first to think about information visualization on the web was Tamara Munzner when she was working at the Geometry Center. Now she is an associate professor in the Department of Computer Science at the University of British Columbia.

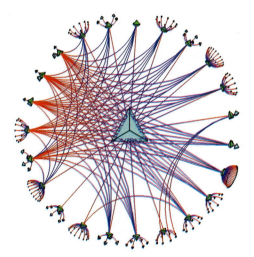

Webpage with its structure based on hyperbolic geometry by Tamara Munzner.

The World Wide Web is so interconnected and huge that it is difficult to establish a mental model of its structure. Hyperbolic geometry offers an elegant way to see the big picture and the interesting details at the same time by using the property that in hyperbolic geometry, the area of a circle grows exponentially with respect to its radius, whereas in Euclidean space the area only grows quadratically. This property allowed visualization of exponentially growing trees representing connections between hyperlinks in a web browser. Practically, this visualization was done on Klein's model of hyperbolic space.[13]

If we want to analyze a large amount of experimental data, for example, from speech recognition, weather forecasts, or changes in economic data, we present these data graphically. Each of these data are dependent on many different parameters, so instead of just simple graphs, known from high school mathematics, we get data in multiple dimensions, and geometrically this means that the graph will be in multidimensional space. We would like to see the shape of the whole graph, but it is impossible. Instead we can analyze the shape by slicing—essentially cutting the graph like a piece of cheese. In the case of cheese, these slices would be planar, but for our data, they can also be hyperbolic planes. Shape representation and shape recognition is another area of collaboration between geometers and computer scientists.

One of the approaches for visualizing and manipulating large data hierarchies is the so-called hyperbolic browser that was inspired by M. C. Escher's woodcut *Circle Limit IV* (Heaven and Hell). The essence of the approach is to lay out the hierarchy on the hyperbolic plane and map this plane onto a circular display region. It is also called "Focus + Context" data display, which means that while users focus on some particular subset of data, they should remain

Picture taken by a "fisheye" camera.

in the context of the entire process. It is like a picture taken by a "fisheye" camera. This camera has an extremely wide field of vision: the center of the image is highly magnified while towards the sides it becomes rapidly smaller. It is the same phenomenon as in the hyperbolic plane representations such as the Poincaré disk model. This approach is used, for example, in bioinformatics to display phylogenetic information.[14]

Network Security

A hyperbolic geometry is useful for network security purposes. Different routes in networks can be represented as graphs. If all data packets are sent via the same, most optimal path, it raises security concerns; data can be intercepted and the full message reconstructed. If the graph is constructed on a negatively curved surface, then because of the properties of this surface, it is possible to have other routes, which are very close to the optimal path but do not coincide with it. This way, different data packages can be sent via different paths, but since the paths are very close, the data arrival time is about the same. This way of sending data is more secure because there are many of these "possibly close" paths to send different parts of the information along.[15]

Music

Western musicians are accustomed to a discrete view of pitch space, corresponding to the chromatic scale playable on the piano. Dmitri Tymoczko presented an ambitious project that characterizes these spaces in great generality and relates the geometry of spaces to the musical behavior of the chords that inhabit them. These spaces are orbifolds

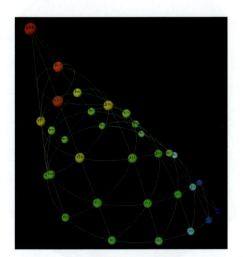

Music chord orbifold image created with Dmitri Tymoczko's Chord-Geometries program.

(which were discussed in Chapter 7) where the local geometry is not Euclidean. Composers in a wide range of styles have exploited the non-Euclidean geometry of these spaces. This is possible because human pitch perception is both logarithmic and periodic. Tymoczko developed a way to geometrically study music and concludes that understanding this non-Euclidean geometry of musical chords will lead to understanding the relation between harmony and counterpoint that may suggest new techniques to contemporary composers.[16]

Art

Lewis Carroll's *Alice in Wonderland* is very well known. But much less known and perhaps never performed is his play, *Euclid and His Modern Rivals*.[17] In this work, Carroll brought back Euclid's ghost to face new challenges in nineteenth-century geometry, including non-Euclidean geometry. His rivals were leading English mathematicians of the day. Carroll even had Euclid's ghost address the proposal by the British Committee for the Improvement of Geometrical Teaching that called for presenting theorems and problems separately.[18]

Ideas deriving from non-Euclidean geometry began to appear in non-mathematical literature in the 1860s. One of the first to popularize new geometric ideas was the great German scientist Hermann von Helmholtz (1821–1894). His articles raised interest about non-Euclidean geometry in literary and artistic circles not only in Germany, but also in France, England, and the United States. The most concentrated non-mathematical manifestation of non-Euclidean geometry was in Paris in the late 1880s and early 1890s. In her book, Linda Henderson discusses the influence of the ideas of non-Euclidean geometry on modern painters.[19]

Modern art was born out of the advanced mathematical and scientific thinking of the time (including hyperbolic geometry). Impressionism and, still more, Cubism brought painting and sculpture much closer to what were the original elements of each: painting as surface design in colors; sculpture as the shaping of volumes to be informed by space.[20] Sculptures of mathematical surfaces with negative curvature are still very popular subjects.

Former physicist, now Professor of Computer Science at U. C. Berkeley, Carlo Séquin is recognized as one of the pioneers of processor design. One of his many interests is computer-aided design (CAD). This professional interest is closely intertwined with his artistic talents. He has created many sculptures, and many of them have negative curvatures. His recent interest is making Escher-type designs.[21]

Costa Surface in a Cube by Carlo Séquin, bronze, 2004 (photo by Carlo Séquin).

Brent Collins' sculptures of surfaces with negative curvature.

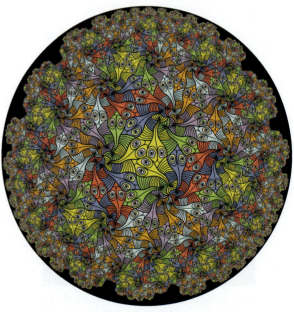

Poincaré FishDish by Carlo Séquin, 2007.

Antipot

Gyroid

Sculptures by Bathsheba Grossman
(photo by Bathsheba Grossman).

Bathsheba Grossman is an artist who also creates her sculptures with CAD, using metal printing technology. She is exploring many geometric forms in her designs, but my favorites, of course, are the ones with negative curvature, such as *Antipot* and *Gyroid*. *Gyroid* is an infinitely connected periodic minimal surface with no straight lines.

In the introduction, I mentioned the crocheted Lorenz manifold by Hinke Osinga and Bernd Krauskopf. In 2006, they met Benjamin Storch—an artist and metal craftsman, who has been researching artistic possibilities of reproducing surfaces of negative curvature. The result was their joint project to create the Lorenz manifold in stainless steel. The surface of this sculpture has positive and negative curvatures.[22]

Benjamin Storch, Bernd Krauskopf, and Hinke Osinga
with their joint project—the sculpture *Manifold*
(photo from B. Krauskopf's personal archive).

The mathematician (author of one of the top ten algorithms of the twentieth century) and sculptor Helaman Ferguson in a 2006 interview[23] admitted that he is in his "negative-Gaussian curvature" phase. Hyperbolic surfaces are often implemented in his sculptures.[24] Granite is Ferguson's favorite medium, and he explains why:

> Mathematics is kind of timeless, so incorporating mathematical themes and ideas into geologically old stone—that's something that has great aesthetic appeal to me.[25]

Poincaré disk models of the hyperbolic plane in Ferguson's sculpture *SYZYGY* are used to represent the planets Venus and Mars. The diameter of Venus is 161 cm, and the diameter of Mars is 174 cm, the average Hamilton College female and male student heights, respectively, according to one of the college's psychology professor's statistics.

SYZYGY by Helaman Ferguson, 2006,
in front of the Science Center at Hamilton College, Clinton, NY,
made from red and beige Texas granite that is one billion years old
(photo by Doug Vasey).

What Else Can Be Done?

Often I hear the question, "Well, it is all very interesting about the hyperbolic plane, but are there really any practical uses besides mathematics, science, and art?" Helaman Ferguson made a poncho for his wife Claire using right-angled pentagon tiling of the hyperbolic plane.

Helaman Ferguson's hyperbolic pentagonal quilt-poncho
(photo by Jeff Weeks).

Romanian industrial designer Radu Comsa saw my crocheted models of the hyperbolic plane and that inspired him to design his Rasta Stool.[26]

These are just some examples of how people come up with their own creative ideas to reproduce the shape of the hyperbolic plane. James Pochert sent me a model he made using a technique called *ply-split braiding*. It is a traditional method for making camel girths and similar bands in India but can also be used to create three-dimensional structures and elaborate patterns.

Rasta Stool (photo by Radu Comsa).

In February 2005, the Institute For Figuring and *Cabinet Magazine* co-organized a talk about crocheting hyperbolic planes in *The Kitchen*, New York City. At the end of the talk, two twin brothers, young artists Trevor and Ryan Oakes, got up and said that they had made their own version of the hyperbolic plane. The audience thought that it was a planted surprise of the show, but I was as much surprised as everyone else. Trevor and Ryan had made their model from pipe cleaners.

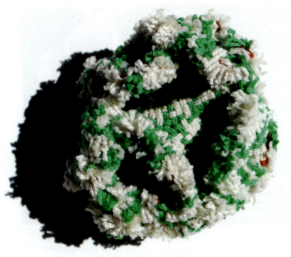

James Pochert used ply-split braiding to create this model.

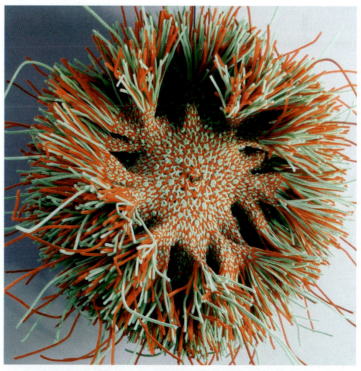

Hyperbolic plane made by Ryan and Trevor Oakes
(photo by Ryan and Trevor Oakes).

I used the crocheted hyperbolic plane pattern to make my own hyperbolic skirt. It starts as the crocheted hyperbolic plane with the initial chain length equal to the size of my waist. Knowing that the number of stitches will increase exponentially (Chapter 6), I used double crochet stitches and an increase ratio of 19 to 20 in order to "slow down" the rapid increase of the length of rows. After I reached the widest part on the hips, I closed my handiwork by not turning around at the end of the row but continuing in a circle (like we did for the pseudosphere in Chapter 6). The length of the skirt really depends on your patience because the rows will continue to get longer and longer. Near the end of the project, it took me the whole evening to complete one row. The project was finished in about three months.

The idea of making models of the hyperbolic plane has grown like the number of stitches in a crocheted model. It started with Bill Thurston's idea to make one from paper annuli. When I learned about it in 1997, topologists and geometers had already been using it for about 20 years. Once crocheted, the hyperbolic first plane was appreciated by students and their instructors. Then, the interest in crocheting the hyperbolic plane spilled out to a much larger audience. Maybe it was because of the soothing effects of this medium or because crochet allows the freedom to explore different forms. After the Institute For Figuring (IFF) commissioned one of my woolen models in late 2005, another artist, Spring Pace, crocheted pieces around it, and the first crocheted coral at the IFF was born.

I am wearing my hyperbolic skirt.

The IFF's first crocheted coral:
my hyperbolic plane model is in the middle
(photo by Christine Wertheim/IFF).

In April 2006, the IFF announced that they started a project called "Hyperbolic Crochet Coral Reef." They contacted other artists and crafters around the world who had already been making their own corals, and also involved hundreds of volunteers who submitted their pieces to the project. You can find out more about this project at www.theiff.org/reef/. This project started from the mathematical idea, but now it has shifted to focus on the important environmental issues that are so essential everywhere in the world.

For me, these crocheting adventures with hyperbolic planes turned out to be quite surprising. When I crocheted my first set of classroom models during the summer of 1997 while camping in the woods in Pennsylvania, I could not possibly think that in June 2006 I would be walking towards 1111 Massachusetts Avenue in Washington, DC, for the opening of the show "Not The Knitting You Know" (curator Binnie Fry), where my crocheted hyperbolic planes were publicly displayed for the first time in a fiber art exhibition. Since then, I have participated in many fiber art shows and have been in several cities in the United States, Belgium, Italy, Latvia, and the United Kingdom. These invitations have challenged me to pay more attention to the colors and materials I choose for my work. For one of the exhibits, I made two pieces from equal amounts of blue and orange cotton yarn, reversing the order. Since the amounts of the yarn are the same, it means that the crocheted areas covered by each of the colors are the same, though visually they look quite different. It made me think about people, how often we think that we are different, we look different, but in our hearts we are the same. The name of my work, *We Are the Same*, grew out of this idea. I was very happy when this message was understood by viewers of the show.

We Are the Same, cotton, 2005.

Paper Models

Appendix

The Hyperbolic Plane from Annuli

This is the paper-and-tape surface originally developed by William Thurston.[1] It is also described in *Experiencing Geometry*.[2] It may be constructed as follows.

Cut out at least ten identical annular strips as in the photo. (An *annulus* is the region between two concentric circles, and we call an *annular strip* a portion of an annulus cut off by an angle from the center of the circles.)

Attach the strips together by attaching the inner circle of one to the outer circle of the other or by attaching them end-to-end (i.e., along the straight ends). (When the straight ends of annular strips are attached together, you get annular strips with increasing "interior" angles, and eventually the angle will be greater than 360°.)

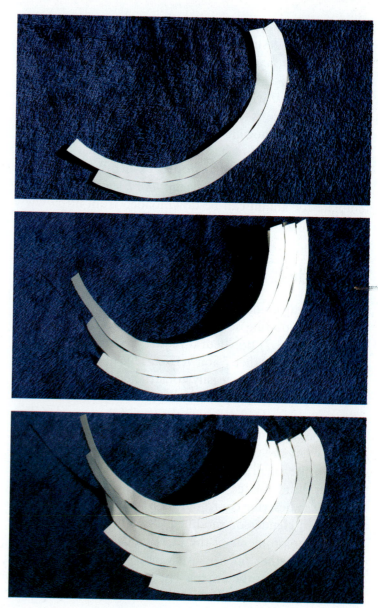

Taping annuli together.

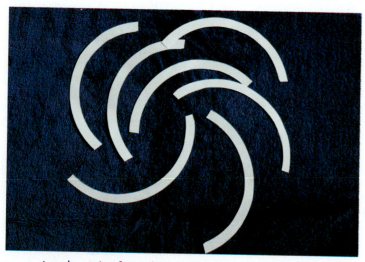

Annular strips for making an annular hyperbolic plane.

The resulting surface is of course only an approximation of the desired surface. Let's call the radius of the smaller circle r_1 and of the bigger circle r_2, and their difference is $\delta = r_2 - r_1$. The actual hyperbolic plane is obtained by letting $\delta \to 0$ while holding the radius fixed. Note that since the surface is constructed the same everywhere (as $\delta \to 0$), it is homogeneous (that is, intrinsically and geometrically, every point has a neighborhood that is isometric to a neighborhood of any other point). We will call the results of this construction the *annular hyperbolic plane*. A template for annuli can be downloaded from http://www.math.cornell.edu/~henderson/ExpGeom/annuli.jpg.

Polyhedral Constructions

A polyhedral model can be constructed from equilateral triangles by putting seven triangles at every vertex.[3] This is called the {3,7} polyhedral model, because triangles (3-gons) are put together 7 at a vertex. This model has the advantage of being more easily constructed than the annular model above; however, one cannot make better and better approximations by decreasing the size of the triangles. This is true because at each vertex the cone angle is 420°, no matter what size the triangles are—whereas the hyperbolic plane in the small looks like the Euclidean plane (360°). Another disadvantage of this polyhedral model is that it is not easy to describe coordinates on it.

David Henderson modified this construction to avoid some of the problems of the {3,7} model by putting seven triangles together only at every other vertex and six triangles together at the others. The precise construction can be described in three different (but, in the end, equivalent) ways:[4]

1. Construct polyhedral annuli as in this figure and then tape them together as with the annular hyperbolic plane.

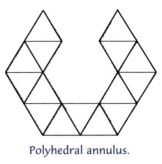

Polyhedral annulus.

2. You can construct two annuli at a time by using the shape in the next figure and taping one of them to the next by joining $a \leftrightarrow A$, $b \leftrightarrow B$, and $c \leftrightarrow C$.

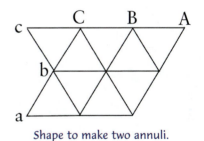

Shape to make two annuli.

3. The quickest way is to start with many strips as pictured in the following figure—these strips can be as long as you wish. Then, add four of the strips together as in the second figure, using five additional triangles (shaded gray). Next, add another strip at every place there is a ver-

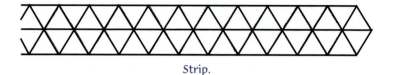

Strip.

Forming the polyhedral annular hyperbolic plane.

Hyperbolic Soccer Ball

Keith Henderson teaches high school geometry. He found the polyhedral model to be very crinkly. He wanted to find a smooth polygonal tiling of the plane and came up with an idea of the *hyperbolic soccer ball*. Essentially, he independently re-discovered tiling {6;6;7} of the hyperbolic plane. Actually, any hyperbolic tiling can be a base for making a model of the hyperbolic plane,[5] but Keith's idea of the hyperbolic soccer ball has worked very well for explaining what the curvature is. This method also allows people to quickly make a paper model of the hyperbolic plane.

Keith's idea was to replace the pentagons in the spherical tiling often seen on soccer balls with heptagons. (See Chapter 1.) Then, the tiling does not close in a sphere but rather makes a model of the hyperbolic plane.

tex with five triangles and a gap (as at the vertices marked with black dots). Every time a strip is added, an additional vertex with seven triangles is formed.

The center of each strip runs perpendicular to each annulus, and you can show that these curves (the center lines of the strip) are all geodesics because they have local reflection symmetry.

Soccer ball (football).

For directions on how to make a hyperbolic soccer ball and a template, see http://www.helpfulservices.net/khenderson/hyperbolicsoccerball.pdf.

I made the model on the facing page out of scraps of fabric by cutting out hexagons and heptagons, and then, instead of real quilting, I just attached them together with iron-on tape.

Another possible hyperbolic tiling is to use four pentagons at each vertex. This method was used in the construction of Helaman Ferguson's quilt-poncho in Chapter 9. For more on this and other tilings of the hyperbolic plane, see the *Science News* article by Ivars Peterson.[6]

Example of art in the making:
hyperbolic paper plane model made in a children's workshop,
Leuven, Belgium, 2006.

Facing page:
Quilt version of hyperbolic soccer ball.

This book would not be possible without the help and support from many people around the world.

The beginning of this project in 2005 was generously supported by a grant from the Institute For Figuring. I would especially like to thank its director, science writer Margaret Wertheim, for her continuous efforts to popularize the beauty of mathematical ideas.

My thanks go to Professor William Thurston for his original idea for a tactile model of the hyperbolic plane and for his helpful suggestions in the writing of this book. I also want to thank Professors Sir Roger Penrose, Robert Osserman, and Maurice Margenstern for sharing their stories and answering my questions.

For advice and assistance with visual materials I would like to thank Manfred Bieber, Imes Chiu, Garry Cobb, Radu Comsa, Douglas Dunham, Helaman Ferguson, Nathaniel Friedman, Paulus Gerdes, Chaim Goodman-Strauss, Bathsheba Grossman, Uwe Grothkopf, Lois Hartwig, Keith Henderson, Jeann Hutchins, Gwen Blakley Kinsler, Bernd Krauskopf, B. J. Licko–Keel, Francis Moon, Elisabeth Mühlhausen, Tamara Munzner, Trevor and Ryan Oakes, Hinke Osinga, James Pochert, Carlo Sequin, Lelde Taimina, Dmitri Tymoczko, Laimdota Upeniece, Stan Wagon, Jeff Weeks, Christine Wertheim, Beverly West, Axel Wittmann, and the Division of Rare and Manuscript Collections at the Cornell University Library.

David Henderson helped me generate ideas for this book and made sure that I persisted with this project all the way, encouraging and supporting me to the end in many ways, including the final typesetting of this book.

The first reviewer of the first draft was my very good friend whose critique and professional advice really helped to shape this book.

Many thanks to Beverly West for all the time she spent this summer carefully reading the manuscript, correcting my chronic language mistakes, and asking very good questions in places where more explanation was needed.

I am grateful to Alice and Klaus Peters for letting me be creative in the writing of this book. Many thanks to Charlotte Henderson (A K Peters) for her patience with all my grammatical mistakes, for good suggestions on how to improve the book, and for generating some computer images. Thanks to Larissa Zaretsky (A K Peters) for the design template and to Kevin Jackson-Mead (A K Peters) for final edits.

Credits for the images used by permission are in the respective captions. The following pictures were obtained from Wikimedia Commons: Pazyryk carpet, All Gizah Pyramids, Parthenon, Ancient building in Uxmal, Antikythera mechanism, Watermill in Belgium, French horn, picture by "fisheye" camera.

All other images are mine or created jointly with my husband, David Henderson.

Thanks to my family, Linda, Lelde, and David, for their patience during the writing of this book.

Daina Taimina
Ithaca, NY
October 2008

Acknowledgments

Many people have an impression, based on years of schooling, that mathematics is an austere and formal subject concerned with complicated and ultimately confusing rules for the manipulation of numbers, symbols, and equations, rather like the preparation of a complicated income tax return, where there are myriad unexplained steps, rules, exceptions, and gotchas.

Good mathematics is quite opposite to this. Mathematics is an art of *human* understanding. Billions of years of evolution have given us many extraordinary capabilities that we ordinarily take for granted—but we deny those capabilities at our peril. In the abstract, the mere act of walking through a room without bumping into other people or things is a far greater accomplishment than the most sophisticated *formal* computation ever done by mathematicians. Computers are far better than humans at formal computations, but humans far surpass current computers at informal and intuitive reasoning.

Our brains are complicated devices, with many specialized modules working behind the scenes to give us an integrated understanding of the world. Mathematical concepts are abstract, so it ends up that there are many different ways they can sit in our brains. A given mathematical concept might be primarily a symbolic equation, a picture, a rhythmic pattern, a short movie—or best of all, an integrated combination of several different representations. The non-symbolic mental models for mathematical concepts are extremely important, but unfortunately, many of them are hard to share.

Mathematics sings when we feel it in our whole brain. People are generally inhibited about even trying to share their personal mental models. People like music, but they are afraid to sing. You only learn to sing by singing.

How do you think of a cat? You probably have mental images, you probably can see in your mind how cats move, you can probably hear in your mind sounds that cats make—perhaps meowing, purring, squeaking, yowling. But can you draw a good picture of a cat, can you give a good description or animation of how a cat moves, can you describe or recreate the sounds a cat makes? It's a lot harder to communicate a cat than to see a cat, and it requires serious cultivation of special talents.

We learn about cats by watching cats; with most mathematical ideas, our culture has provided no comparable shortcuts. Without a good way for whole-brain communication, understanding is denatured.

Non-Euclidean or hyperbolic geometry was a topic of great mystery and confusion for many centuries, as Daina recounts in this book. Insights people may have developed were hard to document, so they crumbled away.

Why do Daina's crochet models have such a great resonance with so many people? It's because they break through the austere, formal stereotype of mathematics and present a path to a whole-brain understanding of a beautiful cluster of simple and significant but rarely understood ideas. The crochet models also break through the stereotype that mathematics is only relevant to traditionally male interests.

These models have a fascination far beyond their visual appearance. As illustrated in the book, there is actually negative curvature and hyperbolic geometry all around us, but people generally see it without seeing it. You will develop an entirely new understanding by actually following the simple instructions and crocheting! The models are deceptively interesting. Perhaps you will come up with your own variations and ideas.

In any case, I hope this book gives you pause for thought and changes your way of thinking about mathematics.

Bill Thurston
Ithaca, NY

Foreword

Contents

Contents

This story started here,
October 1, 1995

Editorial, Sales, and Customer Service Office

A K Peters, Ltd.
888 Worcester Street, Suite 230
Wellesley, MA 02482
www.akpeters.com

Library of Congress Cataloging-in-Publication Data

Taimina, Daina.
  Crocheting adventures with hyperbolic planes / Daina Taimina.
      p. cm.
  Includes bibliographical references and index.
  ISBN 978-1-56881-452-0 (alk. paper)
  1. Geometry, Hyperbolic. 2. Crocheting--History. I. Title.

  QA685.T35 2009
  516.9--dc22

                                                                                    2008038302

Printed in India
13 12 11 10 09   10 9 8 7 6 5 4 3 2 1

Crocheting Adventures with Hyperbolic Planes

Daina Taimiņa

A K Peters, Ltd.
Wellesley, MA

Crocheting Adventures with Hyperbolic Planes

Introduction

1 The previous one had been held 18 years before that, in 1936, in Oslo, Norway; the quadrennial sequence was interrupted because of the war and was renewed in 1950 when the ICM was held in Cambridge, Massachusetts.

2 Miklós Rédei. "'Unsolved Problems in Mathematics': John von Neumann's Address to the International Congress of Mathematicians, Amsterdam, September 2–9, 1954." *The Mathematical Intelligencer* 21 (1999), 7–12.

3 Robert Osserman. "A Hyperbolic Surface in 3-Space." *Proceedings of the American Mathematical Society* 7:1 (1956), 54–58.

4 Doris Schattschneider. "Coxeter and the Artists: Two-Way Inspiration." In *The Coxeter Legacy: Reflections and Projections*, edited by Chandler Davis and Erich W. Ellers, pp. 255–280. Providence, RI: American Mathematical Society and Fields Institute, 2006, p. 262.

5 Lionel S. Penrose and Roger Penrose. "Impossible Objects: A Special Type of Visual Illusion." *British Journal of Psychology* 49:1 (1958), 31–33.

6 Al Seckel. *Masters of Deception: Escher, Dali & the Artists of Optical Illusion.* New York: Sterling Publishing Co., 2004.

7 Roger Penrose. *The Emperor's New Mind: Concerning Computers, Minds, and the Laws of Physics.* Oxford, UK: Oxford University Press, 1990, pp. 418–425.

8 David W. Henderson and Daina Taimina. "Crocheting the Hyperbolic Plane." *The Mathematical Intelligencer* 23 (2001), 17–28.

9 Hinke Osinga and Bernd Krauskopf. "Crocheting the Lorenz Manifold." *The Mathematical Intelligencer* 26:4 (2004), 25–37.

10 Margaret Wertheim. "Crocheting the Hyperbolic Plane: An Interview with David Henderson and Daina Taimina." *Cabinet* 16 (2004). Available at http://www.cabinetmagazine.org/issues/16/crocheting.php.

Chapter 1

1 Consider the same ant on a sphere. Suppose that it plants a post at point P, ties one end of a length of rope to the pole, and ties the other end to itself. She walks away from the pole until the rope is taut and marks that spot. Keeping the rope taut, the ant walks around in a circle until it returns to that mark, measuring how far it walks. If the ant is on a flat surface then the distance of its walk, d, should equal the circumference of a circle with radius equal to the length of its rope, r. If $d < 2\pi r$, then the ant is on a sphere.

2 David Hilbert. "Ueber Flächen von constanter Gaussscher Krümmung." *Transactions of the American Mathematical Society* 2:1 (1901), 87–99. This theorem was improved by E. Holmgren.

3 E. Holmgren. "Sur les surfaces à courbure constante négative." *Les Comptes rendus de l'Académie des sciences* 134 (1902), 740–743.

4 Marc Amsler. "Des surfaces à courbure négative constante dans l'espace à trois dimensions et de leurs singularités." *Mathematische Annalen* 130 (1955), 234–256.

5 This "Nash-Kuiper Theorem" implies that every Riemannian manifold can be isometrically embedded in Euclidean space. Isometrically means that the lengths of the curves are preserved. This result showed that it is possible to visualize any Riemannian manifold as a submanifold in Euclidean space.

6 John Stilwell. *Classical Topology and Combinatorial Group Theory*, Graduate Texts in Mathematics 72. New York: Springer-Verlag, 1993, p. 150.

7 At seven folds it would be about the thickness of the notebook, at 17 about the height of a two-story house. If you live in a high-rise apartment building, then keep going!

Chapter 2

1 Michele Emmer (editor). *The Visual Mind II.* Cambridge, MA: MIT Press, 2005, p. 78.

2 Henri Poincaré. *The Value of Science.* New York: Dover, 1958, p. 37.

[3] The fold demonstrates bilateral (mirror) symmetry of the fold line, which is a determining characteristic of straightness.

[4] A. T. Olson. *Mathematics though Paper Folding*. Reston, VA: NCTM, 1975.

[5] Letter from Carl Freidrich Gauss to Taurinus in 1824; quoted from: *Gauss Werke Bd. VIII*. Göttingen: Königliche Gesellschaft der Wissenschaften, 1900.

[6] David E. Rowe. "Euclidean Geometry and Physical Space." *The Mathematical Intelligencer* 28:2 (2006), 51–60.

[7] David W. Henderson and Daina Taimina. *Experiencing Geometry: Euclidean and Non-Euclidean with History*, third edition. Upper Saddle River, NJ: Prentice Hall, 2005, Chapter 5.

[8] It is important to note here that we cannot use a normal compass to draw these circles. This is because we want the intrinsic radii (the radii as measured along the surface) of the three circles to be the same. If we used a compass then the intrinsic radius on the sphere would be more than the radius on the paper. It is an interesting fact that if you use a compass to mark a circle on the paper and on the sphere, these two circles will have the same area, as can be determined by the interested reader with knowledge of trigonometry and spherical geometry.

Chapter 3

[1] H. Graham Flegg. *From Geometry to Topology*. New York: Dover, 2001, p. 168.

[2] Owen Jones. *The Grammar of Ornament*. Lincoln's Inn Fields, London, Day&Son, 1856; first American edition 2001 by DK Publishing, Inc., p. 31.

[3] O. B. Duane. *Celtic Art*. London: Brockhampton Press, 1996.

[4] Mathematicians are now recognizing these drawings as *Gaussian graphs*: a planar map that can be drawn by a simple closed curve with no more than two pieces of the curve intersecting in the same point.

[5] Paulus Gerder. *Geometry from Africa: Mathematical and Educational Explorations*. Washington, DC: Mathematical Association of America, 1999, Chapter 4.

[6] The Villa is now a UNESCO World Heritage site. It was constructed in the early fourth century on the remains of an older villa. It was destroyed and abandoned in the twelfth century and rediscovered in the seventeenth century when some peasants noticed numerous walls sticking out of the ground. From that moment, many unauthorized excavations occurred, searching for valuable objects. Only in 1955 did some serious restoration and preservation begin. The mosaics are the prime reason for the fame of the villa.

[7] Keith Critchlow. *Islamic Patterns: An Analytical and Cosmological Approach*. Rochester, VT: Inner Traditions, 1999.

[8] Linda Henderson. *The Fourth Dimension and Non-Euclidean Geometry in Modern Art*. Princeton, NJ: Princeton University Press, 1983, p. 98.

[9] A. Seidenberg. "The Ritual Origin of Geometry." *Archive for History of Exact Sciences* 1(1962), 488–527.

[10] Ibid, p. 492.

[11] Reginald Laubin and Gladys Laubin. *Indian Dances of North America*. Norman, OK: University of Oklahoma Press, 1989; Peter Nabokov and Robert Easton. *Native American Architecture*. Oxford, UK: Oxford University Press, 1989.

[12] John G. Neihardt. *Black Elk Speaks*. Lincoln, NE: University of Nebraska Press, 1961.

[13] I can suggest, for example: Priya Hemenway. *Divine Proportion: Phi in Art, Nature, and Science*. New York: Sterling Publishing, 2005; or Hans Walser. *The Golden Section*. Washington, DC: The Mathematical Association of America, 2001.

[14] Euclid's five postulates are the basic rules (axioms) that determine Euclidean geometry; for example, the first one states that a straight line can be drawn between any two points.

[15] Marcia Asher. *Mathematics Elsewhere: An Exploration of Ideas Across Cultures*. Princeton, NJ: Princeton University Press, 2002.

[16] A compass and unmarked straightedge sequence is a sequence of points, lines, and circles that starts with two distinct given points and is such that each of the other points in this sequence is the intersection of lines or circles that occur before it in the sequence; each circle has its center and one point on the bound-

ary occurring before it in the sequence, and each line contains two distinct points that occur before it in the sequence. See David W. Henderson and Daina Taimina. *Experiencing Geometry: Euclidean and Non-Euclidean with History*, third edition. Upper Saddle River, NJ: Prentice Hall, 2005, pp. 213–216.

[17] Tony Freeth, Alexander Jones, John M. Steel, and Yanis Bitsakis. "Calendars with Olympiad Display and Eclipse Prediction on the Antikythera Mechanism." *Nature* 454 (2008), 614–617.

[18] J. Fauvel and J. Gray (editors). *The History of Mathematics: A Reader*. Milton Keynes, UK: Open University, 1987, p. 175.

Chapter 4

[1] Kristin Bühler. "Classification of Basic Textile Techniques." *Ciba Review* 63 (1948), 2289–2320.

[2] Heinz Edgar Kiewe. *The Sacred History of Knitting*. Oxford: Art Needlework Industries Ltd, 1967, pp. 21–22.

[3] Mary Konior. *Heritage Crochet: An Analysis*. London: Dryad Press, 1987.

[4] Heinz Edgar Kiewe. *The Sacred History of Knitting*. Oxford: Art Needlework Industries Ltd, 1967.

[5] Ibid, p. 27–28.

[6] Kristin Bühler. "Classification of Basic Textile Techniques." *Ciba Review* 63 (1948), 2289–2320.

[7] Lis Paludan. *Crochet: History & Technique*. Loveland, CO: Interweave Press, 1995.

[8] Ibid.

[9] Annie Potter. *Living Mystery: The International Art & History of Crochet*. Big Sandy, TX: AJ Publishing International, 1990.

[10] Ibid, p. 12.

[11] Ibid, p. 14.

[12] Mary Konior. *Heritage Crochet: An Analysis*. London: Dryad Press, 1987, p. 6.

[13] Clinton D. MacKenzie. *New Design in Crochet*. New York: Van Nostrand Reinhold Company, 1972.

[14] Elisabeth Mühlhausen. "Riemann Surface—Crocheted in Four Colors." *The Mathematical Intelligencer* 15:3 (1993), 49–53.

Chapter 5

[1] John Fauvel and Jeremy Gray (editors). *The History of Mathematic: A Reader*. Milton Keynes, UK: The Open University, 1987.

[2] Proclus (translated by Glenn R. Morrow). *Proclus: A Commentary on the First Book of Euclid's Elements*. Princeton, NJ: Princeton University Press, 1970.

[3] Hyperbolic trigonometric functions are analogs to ordinary trigonometric functions. The difference is that ordinary trigonometric functions are defined on a unit circle so that $\sin^2 \theta + \cos^2 \theta = 1$, but hyperbolic trigonometric functions are defined on a hyperbola so that $\cosh^2 \theta - \sinh^2 \theta = 1$.

[4] David W. Henderson and Daina Taimina. *Experiencing Geometry: Euclidean and Non-Euclidean with History*, third edition. Upper Saddle River, NJ: Prentice Hall, 2005.

[5] Jeremy J. Gray, *Janos Bolyai, Non-Euclidean Geometry and the Nature of Space*. Cambridge, MA: The MIT Press, 2004.

[6] Jeremy Gray, *Ideas of Space: Euclidean, non-Euclidean, and Relativistic*. Oxford: Clarendon Press, 1990.

[7] G. Waldo Dunnigton. *Carl Friedrich Gauss, Titan of Science*. New York: Hafner, 1955; reprinted by the MAA in 2004, p. 44.

[8] John Fauvel and Jeremy Gray (editors). *The History of Mathematic: A Reader*. Milton Keynes, UK: The Open University, 1987.

[9] Jeremy J. Gray, *Janos Bolyai, Non-Euclidean Geometry and the Nature of Space*. Cambridge, MA: The MIT Press, 2004.

[10] Ibid, p. 51.

[11] For the complete letter, see ibid, p. 54.

[12] Ibid, p. 53.

[15] David W. Henderson and Daina Taimina. "Crocheting the Hyperbolic Plane." *The Mathematical Intelligencer* 23 (2001), 17–28.

[16] *Ruffled Doilies*, The Spool Cotton Company, distributor J&P Coats, New York, Book No. 253, 1949.

13 Ibid, p. 82.

14 Marcel Berger. *Panoramic View of Riemannian Geometry.* Berlin: Springer, 2007.

15 Felix Klein. *Vergleichende Betrachtungen über neuere geometrische Forschungen.* Erlangen: Verlag von Andreas Deichert, 1872.

16 Felix Klein (translated by M. W. Haskell). "A Comparative Review of Recent Researches in Geometry." *Bulletin of the New York Mathematical Society* 2 (1892–1893), 215–249.

17 For more about this topic, see: David W. Henderson and Daina Taimina. *Experiencing Geometry: Euclidean and Non-Euclidean with History,* third edition. Upper Saddle River, NJ: Prentice Hall, 2005; and Donal O'Shea. *Poincaré Conjecture: In Search of the Shape of the Universe.* New York: Walker and Co., 2007.

18 Linda Dalrymple Henderson. *The Fourth Dimension and Non-Euclidean Geometry in Modern Art.* Princeton, NJ: Princeton University Press, 1983, p. 100.

19 Ibid.

Chapter 6

1 J. Dinsdale. "Horn Loudspeaker Design." *Wireless World* (March 1974), 19–24.

2 Steven Battersby. "Big Bang Glows Hints at Funnel-Shaped Universe." *New Scientist,* April 15, 2004. Available at http://www.newscientist.com/article.ns?id=dn4879.

3 The basic theory behind this can be found in David W. Henderson. *Differential Geometry.* Upper Saddle River, NJ: Prentice Hall, 2005.

4 See, for example, the online calculator http://www.mathsisfun.com/scientific-calculator.html.

Chapter 7

1 In his semi-autobiographical work *Down and Out in Paris and London,* first published in 1933.

2 Paraphrased from Reuben Hersh. "Mathematics has a front and a back." *Synthese* 88 (1991), 127–133.

3 William Byers. *How Mathematicians Think: Using Ambiguity, Contradiction, and Paradox to Create Mathematics.* Princeton, NJ: Princeton University Press, 2007, p. 25.

4 Commonly, "manifold" refers to a "topological manifold," but these differ from the "geometric manifolds" that are discussed in this book. The focus is on geometric manifolds in order to investigate geometric properties.

5 William P. Thurston. *Three-Dimensional Geometry and Topology,* Vol. 1. Princeton, NJ: Princeton University Press, 1997.

6 Cat Bordhi. *A Treasury of Magical Knitting.* Friday Harbor, WA: Passing Paw Press, 2004; Cat Brodhi. *A Second Treasury of Magical Knitting.* Friday Harbor, WA: Passing Paw Press, 2005.

7 Amy F. Szczepański. "Quilted Möbius Band." In *Making Mathematics with Needlework: Ten Papers and Ten Projects,* edited by sarah-marie belcastro and Carolyn Yackel. Wellesley, MA: A K Peters, 2008, pp. 13–28.

8 Clifford Pickover. *The Möbius Strip: Dr. August Möbius's Marvelous Band in Mathematics, Games, Literature, Art, Technology, and Cosmology.* New York: Thunder's Mouth Press, 2006.

9 Susan Goldstine. "Fortunatus's Purse." In *Making Mathematics with Needlework: Ten Papers and Ten Projects,* edited by sarah-marie belcastro and Carolyn Yackel. Wellesley, MA: A K Peters, 2008, pp. 105–117.

10 John H. Conway, Heidi Burgiel, and Chaim Goodman-Strauss. *The Symmetries of Things.* Wellesley, MA: A K Peters, 2008; Chapter 9.

11 For example, in William P. Thurston. *Three-Dimensional Geometry and Topology,* Vol. 1. Princeton, NJ: Princeton University Press, 1997, p. 16; or a picture in George K. Francis. *A Topological Picturebook.* Berlin: Springer, 1987, pp. 35, 127; or in Jeffery Weeks. *The Shape of Space.* New York: Marcel Dekker, 2002, p. 235.

12 See David W. Henderson and Daina Taimina. *Experiencing Geometry: Euclidean and Non-Euclidean with History,* third

edition. Upper Saddle River, NJ: Prentice Hall, 2005, Chapter 7.

13 For example, see Marcel Bergen. *A Panoramic View of Riemannian Geometry.* New York: Springer, 2003, p. 157.

14 William Thurston. Foreword to *Teichmüller Theory and Applications to Geometry, Topology and Dynamics,* by J. Hubbard. Ithaca, NY: Matrix Editions, 2006.

15 William P. Thurston. *Three-Dimensional Geometry and Topology,* Vol. 1. Princeton, NJ: Princeton University Press, 1997, p. 110.

16 See John G. Ratcliffe. *Foundations of Hyperbolic Manifolds.* New York: Springer, 1994, p. 147.

17 Jeffrey Weeks. *The Shape of Space,* second edition. New York: Marcel Dekker, Inc., 2002.

18 David W. Henderson and Daina Taimina. *Experiencing Geometry: Euclidean and Non-Euclidean with History,* third edition. Upper Saddle River, NJ: Prentice Hall, 2005.

Chapter 8

1 Robert Osserman. *A Survey of the Minimal Surfaces.* New York: Dover, 1969.

2 J. Maxfield and J. L. Coolidge. *A History of Geometrical Methods.* New York: Dover, 2003, p. 329. (Originally printed in 1940.)

Chapter 9

1 http://www.claymath.org/millennium/P_vs_NP/

2 Maurice Margenstern and Kenichi Morita. "NP Problems are Tractable in the Space of Cellular Automata in the Hyperbolic Plane." *Theoretical Computer Science* 259 (2001), 99–128.

3 Maurice Margenstern. "Cellular Automata and Combinatoric Tilings in Hyperbolic Spaces: A Survey." In *Discrete Mathematics and Theoretical Computer Science: 4th International Conference, DMTCS 2003, Dijon, France, July 7–12, 2003, Proceedings,* Lecture Notes in Computer Science 2731. Berlin: Springer, 2003, pp. 48–72.

4 Barry Mazur. "Number Theory as Gadfly." *The American Mathematical Monthly* 98:7 (1991), 593–610.

5 Lakshminarayanan Mahadevan. Private communication, summer 2006, and Jonathon Shaw. "Physics of the Familiar." *Harvard Magazine* 110:4 (2008). Available at http://harvardmagazine.com/2008/03/p-the-physics-of-the-famil.html.

6 Yael Klein, Efi Efrati, and Eran Sharon. "Shaping of Elastic Sheets by Prescription of Non-Euclidean Metrics." *Science* 315 (2007), 1116–1119.

7 Reinhard Nesper and Stefano Leoni. "On Tilings and Patterns on Hyperbolic Surfaces and Their Relation to Structural Chemistry." *ChemPhysChem* 2:7 (2001), 413–422.

8 *Mathematics Illuminated,* 13-part TV series produced by Annenberg Media, 2008, Part 8, "Geometries Beyond Euclid."

9 J. Richard Gott III. "Topology and the Universe." *Classical and Quantum Gravity* 15 (1998), 2713–2719.

10 More about this in Jeffrey Weeks. *The Shape of Space.* New York: Marcel Dekker, Inc., 2002; and Jean-Pierre Luminet, *The Wraparound Universe.* Wellesley, MA: A K Peters, 2008.

11 Eduardo Brandoni da Silva, Marcelo Firer, Sueli R. Costa, and Reginaldo Palazzo, Jr. "Signal Constellations in the Hyperbolic Plane: A Proposal for New Communication Systems." *Journal of the Franklin Institute* 343 (2006), 69–82.

12 J. M. Vigoureux and R. Giust. "The Use of the Hyperbolic Plane in Studies of Multilayers." *Optics Communications* 186 (2000), 231–236.

13 Tamara Munzner and Paul Burchard. *Visualizing the Structure of the World Wide Web in 3D Hyperbolic Space.* New York: ACM Press, 1995.

14 John Lamping, Ramana Rao, and Peter Pirolli. "A Focus+Context Technique Based on Hyperbolic Geometry for Visualizing Large Hierarchies." In *Proceedings of the SIGCHI Conference on Human Factors in Computing Systems.* New York: ACM Press, 1995, pp. 401–408.

15 E. A. Jonckheere and P. Lohsoonthorn. "A Hyperbolic Geometry Approach to Multipath Routing." Paper presented at the 10th Mediterranean Conference on Control and Automation (MED 2002), Lisbon, Portugal, July 9–12, 2002.

[16] Dmitri Tymoczko. "The Geometry of Musical Chords." *Science* 313:5783 (2006), 72–74.

[17] Charles Lutwidge Dodgson. *Euclid and His Modern Rivals*, second edition. London: Macmillan and Co., 1855. Available online from Cornell Library Digital Math Books Collection, http://astech.library.cornell.edu/ast/math/additional/Digital-Books.cfm.

[18] David E. Rowe. "Euclidean Geometry and Physical Space." *Mathematical Intelligencer* 28:2 (2006), 51–60.

[19] Linda Dalrymple Henderson. *The Fourth Dimension and Non-Euclidean Geometry in Modern Art*. Princeton, NJ: Princeton University Press, 1983.

[20] Suzanne Küchler. "Why Knot? Towards a Theory of Art and Mathematics." In *Beyond Aesthetics: Art and the Technologies of Enchantment*, edited by Christopher Pinney and Nicholas Thomas. Oxford: Berg, 2001, pp. 57–78.

[21] Carlo Séquin. "Symmetric Embedding of Locally Regular Hyperbolic Tiling." Paper presented at Bridges Conference for Mathematical Connections in Art, Music, and Science, San Sebastian, Spain, July 24–27, 2007; available at www.cs.berkeley.edu/~sequin/PAPERS/2007_Bridges_HyperbolicTiles.pdf.

[22] Bernd Krauskopf, Hinke M. Osinga, Benjamin Storch. "The sculpture *Manifold*: A Band from a Surface, a Surface from a band". *Proceedings of 2008 Bridges Leeuwarden*. Hertfordshire, UK: Tarquin Publishing, 2008.

[23] Katherine Unger. "Carving His Own Unique Niche, in Symbols and Stone." *Science* 314:5798 (2006), 412–413.

[24] H. Ferguson, A. Rockwood, and J. Cox. "Topological Design of Sculptured Surfaces." *Computer Graphics* 26:2 (1992), 149–156; Claire Ferguson and Helaman Ferguson. *Mathematics in Stone and Bronze*. Erie, PA: Meridian Creative Group, 1994.

[25] Katherine Unger. "Carving His Own Unique Niche, in Symbols and Stone." *Science* 314:5798 (2006), 412–413.

[26] http://www.raducomsa.ro/furniture/full_rs.html.

Appendix

[1] William P. Thurston. *Three-Dimensional Geometry and Topology*, Vol. 1. Princeton, NJ: Princeton University Press, 1997, p. 50.

[2] David W. Henderson and Daina Taimina. *Experiencing Geometry: Euclidean and Non-Euclidean with History*, third edition. Upper Saddle River, NJ: Prentice Hall, 2005.

[3] Jeffrey Weeks. *The Shape of Space*. New York: Marcel Dekker, 2002, p. 151.

[4] The drawings here first appeared in *Experiencing Geometry* and are used here by permission.

[5] See the nice gallery of hyperbolic tilings by Bernie Freidin at http://bork.hampshire.edu/~bernie/hyper/.

[6] Ivars Peterson. "Math Trek: Hyperbolic Five." *Science News Online* August 30, 2003, available at http://63.240.200.111/articles/20030830/mathtrek.asp.

Index